T0137213

Studies in Big Data

Volume 115

Series Editor

Janusz Kacprzyk, Polish Academy of Sciences, Warsaw, Poland

The series "Studies in Big Data" (SBD) publishes new developments and advances in the various areas of Big Data- quickly and with a high quality. The intent is to cover the theory, research, development, and applications of Big Data, as embedded in the fields of engineering, computer science, physics, economics and life sciences. The books of the series refer to the analysis and understanding of large, complex, and/or distributed data sets generated from recent digital sources coming from sensors or other physical instruments as well as simulations, crowd sourcing, social networks or other internet transactions, such as emails or video click streams and other. The series contains monographs, lecture notes and edited volumes in Big Data spanning the areas of computational intelligence including neural networks, evolutionary computation, soft computing, fuzzy systems, as well as artificial intelligence, data mining, modern statistics and Operations research, as well as self-organizing systems. Of particular value to both the contributors and the readership are the short publication timeframe and the world-wide distribution, which enable both wide and rapid dissemination of research output.

The books of this series are reviewed in a single blind peer review process.

Indexed by SCOPUS, EI Compendex, SCIMAGO and zbMATH.

All books published in the series are submitted for consideration in Web of Science.

Julio C. Urenda · Vladik Kreinovich

Algebraic Approach to Data Processing

Techniques and Applications

 Springer

Julio C. Urenda
Department of Computer Science
University of Texas at El Paso
El Paso, TX, USA

Vladik Kreinovich
Department of Computer Science
University of Texas at El Paso
El Paso, TX, USA

ISSN 2197-6503 ISSN 2197-6511 (electronic)
Studies in Big Data
ISBN 978-3-031-16782-9 ISBN 978-3-031-16780-5 (eBook)
https://doi.org/10.1007/978-3-031-16780-5

This Springer imprint is published by the registered company Springer Nature Switzerland AG
The registered company address is: Gewerbestrasse 11, 6330 Cham, Switzerland

Preface

In many areas of human knowledge, symmetries and invariances play an important role. In fundamental physics, starting with Relativity Theory, new physical theories have been formulated in terms of invariances and of the corresponding transformation groups—i.e., in terms what a mathematician would call an algebraic approach. In engineering, devices like wind tunnels, which are based on scale-invariance, enable us to test smaller-scale models of the actual designs. In biological sciences, symmetries and invariances are extremely important in analyzing the shape and functioning of living beings, from mammals to viruses. Invariance and symmetry—in the form of fairness—are an extremely important topic in social sciences.

Because of ubiquity of invariances, it is reasonable to take them into account when processing data corresponding to different domains. In this book, we show, on examples from various application domains—physics, engineering, medicine, economics, social sciences, education, even mathematics itself—that the algebraic approach is indeed very helpful in data processing, both in providing theoretical justifications for heuristic techniques and in coming up with new more efficient data processing methods. We also show that algebraic approach is helpful not only in specific applications, but also in analyzing and developing computational methods leading to these applications—methods ranging from deep learning to fuzzy and probabilistic techniques to other promising techniques such as DNA computing.

El Paso, USA

Julio C. Urenda
Vladik Kreinovich

Contents

Chapter 1
Introduction

1.1 What Is Data Processing and Why Do We Need It?

One of the main objectives of science and engineering is:

- to predict the future state of the world—i.e., the future values of the quantities that characterize this state, and
- to find the way to make this future state better—by finding the appropriate values of the parameters of the gadgets and of their controls.

To make these predictions—and to find the appropriate designs and controls—we need to take into account the current state of the world and the knowledge of how the world changes. The values of the quantities describing the current state of the world come from measurements or from expert estimates. Based on our knowledge of how the world changes, we design algorithms for processing the resulting data. This, in a nutshell, is what data processing is about.

1.2 Why Algebraic Approach?

How do we gain knowledge about the world? For example, how did we learn that if we drop an object, it will fall with the acceleration of 9.81 m/s^2? Well, the scientists dropped an object once, and observed this fall. Then they moved to a different location and repeated the same experiment—and got the exact same result. Then they turned by an angle—and also got the same result. After several such experiments, they concluded that the result of this experiment does not change if we move to a different location or turn by some angle. In other words, they concluded that this process is *invariant* with respect to shifts and rotations.

© The Author(s), under exclusive license to Springer Nature Switzerland AG 2022
J. C. Urenda and V. Kreinovich, *Algebraic Approach to Data Processing*,
Studies in Big Data 115, https://doi.org/10.1007/978-3-031-16780-5_1

In other cases, other transformations are appropriate. For example, the whole idea of a wind tunnel—in which smaller-size airplane models used to be tested—is that the corresponding processes do not change if we re-scale the objects. In electrodynamics, all interactions remain the same if we replace all positive charges with negative ones, and vice versa. According to Special Relativity theory, processes do not change if everything starts moving with a constant speed, etc.

In all these cases, we have some transformations with respect to which processes are invariant. In mathematics, studying the classes of such transformation is classified as part of *algebra*. From this viewpoint, algebraic approach to designing (and optimizing) data processing algorithms is a very natural idea.

In this book, we describe applications of this idea to various aspects of algorithmics.

Comment. An important particular case of invariance is invariance with respect to geometric transformations. Such invariances are known as *symmetries*. Based of this, all invariances are often called *symmetries*. Under this name, they are one of the most important tools in physics (see, e.g., [1, 2]) to the extent that nowadays, new physical theories are usually formulated not in terms of differential equations—as in Newton's times—but in terms of the corresponding symmetries.

1.3   What We Do in This Book: An Overview

We start with the general analysis of the situation:

- what are the most natural and the most frequent transformations (Chap. 2),
- which functions and which families of functions are invariant with respect to these transformations (Chap. 3), and
- what is the general relation between invariance and optimality (Chap. 4).

Once we have presented the corresponding results, we describe applications of algebraic approach—applications both to providing theoretical justifications for the existing heuristic application techniques and to developing new techniques. In particular, we list the following applications:

- to dynamic systems in general—in Chap. 5;
- to physics—in Chaps. 6 and 7;
- to engineering—in Chaps. 8 and 9;
- to medicine—in Chaps. 10–15;
- to economics—in Chaps. 16–22;
- to social sciences—in Chap. 23;
- to education—in Chap. 24–25; and
- to mathematics—in Chap. 26.

In addition to specific *applications*, we also apply algebraic techniques to *computational methods* leading to such applications. At present, the most promising data processing techniques are techniques corresponding to machine learning, especially to neural networks, in particular, to deep neural networks. These techniques are very successful, but they are not perfect.

One of the problems is that these techniques are largely heuristic, many of their features lack a solid theoretical foundation—a foundation that would increase our trust in these techniques. In Chap. 27, we use algebraic approach to provide a justification for some of these features—and we show that this justification also explains a successful heuristic modification of some of these features. Another modification is justified in Chap. 28.

Yet another problem of deep learning is that it is a black box, its results do not come with any explanations. If we could add some natural-language explanations, that would make these results more convincing and thus, more acceptable. Need for such explanations—i.e., for understandability of the results—leads to considering fuzzy techniques, see Chap. 29. In Chap. 30, we show that this leads to a new justification for rectified linear activation functions which are so successfully used in deep learning; we also explain which versions of fuzzy techniques is computationally the fastest. In Chap. 31, we analyze which versions are the best if we use different criteria. Two auxiliary fuzzy-related results are described in Chaps. 32 and 33: which degrees to use and how to explain some features of commonsense reasoning.

A known alternative to fuzzy techniques are probabilistic techniques. They work perfectly well when we have the full information about the corresponding probabilities, but in many practical situations, we do not have this information. In this case, as we show in Chaps. 34 and 35, algebraic approach can help select the corresponding probability distributions; limitations of this selection are described in Chap. 36. Finally, in Chap. 37, we explore data processing techniques beyond neural and fuzzy: namely, DNA computing.

1.4 Thanks

Many thanks to our colleagues Martine Ceberio, and Piotr Wojciechowski for their help. Last but not the least, many thanks for our families, our spouses Esmeralda and Olga and our sons Alexander and Misha for their love and patience.

References

1. Feynman, R., Leighton, R., Sands, M.: The Feynman Lectures on Physics. Addison Wesley, Boston, Massachusetts (2005)
2. Thorne, K.S., Blandford, R.D.: Modern Classical Physics: Optics, Fluids, Plasmas, Elasticity, Relativity, and Statistical Physics. Princeton University Press, Princeton, New Jersey (2017)

Chapter 2
What Are the Most Natural and the Most Frequent Transformations

2.1 Main Idea: Numerical Values Change When We Change a Measuring Unit and/or Starting Point

In data processing, we deal with numerical values of different physical quantities. Computers just treat these values as numbers, but from the physical viewpoint, it is important to understand that the numerical values are not absolute: they change if we change the measuring unit and/or the starting point for measuring the corresponding quantity.

2.2 Scaling Transformations

Let us first analyze what happens if we change the measuring unit. For example, we can measure a person's height in meters or in centimeters. The same height of 1.7 m, when described in centimeters, becomes 170 cm. In general, if we replace the original measuring unit with a new unit which is λ times smaller, then for each physical quantity, instead of the original numerical value x, we get a new numerical value $\lambda \cdot x$—while the actual quantity remains the same. The corresponding transformation $x \to \lambda \cdot x$ is known as *scaling*.

2.3 Shifts

For some physical quantities, e.g., for time or temperature, the numerical value also depends on the starting point. For example, we can measure the time by counting how much time has passed during the flight—i.e., by using the flight start time as

© The Author(s), under exclusive license to Springer Nature Switzerland AG 2022
J. C. Urenda and V. Kreinovich, *Algebraic Approach to Data Processing*,
Studies in Big Data 115, https://doi.org/10.1007/978-3-031-16780-5_2

a starting point. Alternatively, we can use the usual calendar time, in which Year 0 is the starting point. In general, if we replace the original starting point with the new one which is x_0 units earlier, than each original numerical value x is replaced by a new numerical value $x + x_0$. The corresponding transformation $x \rightarrow x + x_0$ is known as a *shift*.

2.4   Linear Transformations

In general, if we change both the measuring unit and the starting point, we get a linear transformation: from the original value x, we get to $\lambda \cdot x + x_0$. A usual example of such a transformation is a transition from Celsius to Fahrenheit temperature scales: if we know the temperature t_C is Celsius, then the Fahrenheit temperature t_F is equal to $t_F = 1.8 \cdot t_C + 32$.

2.5   Geometric Transformations

In the previous text, we only considered transformations that transform the value of a single variable. In addition to such transformations, we also have natural transformation that transform two or more variables. For example, a planar rotation transforms the coordinates (x, y) into new coordinates (X, Y) which are related to the original ones by linear formulas.

In addition to rotations, it sometimes makes sense to consider more general linear transformations $x_i \rightarrow a_i + \sum_{j=1}^{n} a_{ij} \cdot x_j$ known as *affine* transformations.

2.6   Beyond Linear Transformations

In the previous text, we considered linear transformations between different scales. In some cases, the relation between different scales is nonlinear. For example, we can measure the earthquake energy is Joules (i.e., in the usual scale) or in a logarithmic (Richter) scale.

The possibility of nonlinear transformations raises a natural question: what are the natural transformations between different scales?

- First, as we have argued in the previous text, all linear transformations are natural.
- Second, if we have a natural transformation $f(x)$ from scale A to another B, then the inverse transformation $f^{-1}(x)$ from scale B to scale A should also be natural.

- Third, if $f(x)$ and $g(x)$ are natural scale transformation, then we can apply first $g(x)$ to get $y = g(x)$ and then f to get $f(y) = f(g(x))$. Thus, the composition $f(g(x))$ of two natural transformations should also be natural.

In mathematical terms, the class of transformations that contain an identity mapping $f(x) = x$ and that satisfies the second and third properties is called a *transformation group*. In these terms, the above properties can be reformulated as follows: the class T of natural transformations is a transformation group that contains all linear transformations.

We also need to take into account that in a computer, at any given moment of time, we can only store the values of finitely many parameters. Thus, the transformations from the desired transformation group T should be determined by a finite number of parameters. In mathematical terms, the smallest number of parameters needed to describe a family is known as the *dimension* of this family—just like the fact that we need 3 coordinates to describe any point in space means that the physical space is 3-dimensional. In these terms, the transformation group T must be finite-dimensional.

Interestingly, the above requirements uniquely determine the class of all possible natural transformation. This result can be traced back to Norbert Wiener, the father of cybernetics. In his seminal book *Cybernetics* [1] that started this research area, he noticed that when we approach an object form afar, our perception of this object goes through several distinct phases:

- first, we see a blob; this means that at a large distance, we cannot distinguish between images obtained each other by all possible continuous transformations; in other words, this phase corresponds to the group of all possible continuous transformations; transformations;
- as we get closer, we start distinguishing angular parts from smooth parts, but still cannot compare sizes; this corresponds to the group of all projective transformations;
- after that, we become able to detect parallel lines; this corresponds to the group of all transformations that preserve parallel lines—i.e., to the group of all linear (= affine) transformations;
- when we get even closer, we become able to detect the shapes, but we still cannot distinguish between larger objects that are further away and smaller objects which are closer—our binocular vision (that enables us to make this distinction) only starts working at shorter distances; this corresponds to the group of all homotheties;
- finally, as we get much closer, we see the exact shapes and sizes; this means that only the identity transformation remains.

Wiener argued that there are no other transformation groups—since if there were other transformation groups, after billions years of evolutions, we would use them. In precise terms, he conjectured that the only two finite-dimensional transformation groups that contain all linear transformations are the groups of all linear transformations and the group of all projective transformations.

Interestingly, this was proven to be true, so we need to consider projective transformations. In particular, for transformations of the real line, projective transformations are simply fractional-linear transformations; see, e.g., [2, 3] and references therein:

$$f(x) = \frac{a \cdot x + b}{c \cdot x + d}.$$

2.7 Permutations

In the discrete case, it is often natural to consider *permutations* $\pi : \{1, \ldots, n\} \rightarrow \{1, \ldots, n\}$. For example, if we are considering solutions to economic situations involving several participants, then fairness means, in particular, that the solution should not depend on the order in which these participants are presented, i.e., that this solution should be invariant with respect to all possible permutations.

References

1. Wiener, N.: Cybernetics: Or Control and Communication in the Animal and the Machine. MIT Press, Cambridge, Massachisetts (1948)
2. Kreinovich, V., Quintana, C.: Neural networks: what non-linearity to choose? In: Proceedings of the 4th University of New Brunswick AI Workshop, Fredericton, New Brunswick, Canada (1991) pp. 627–637
3. Nguyen, H.T., Kreinovich, V.: Applications of Continuous Mathematics to Computer Science. Kluwer, Dordrecht (1997)

Chapter 3
Which Functions and Which Families of Functions Are Invariant

3.1 Why Do We Need Invariant Functions

As we have mentioned earlier, our main objective is to study the relation between several quantities. In computational terms, this relation corresponds to an algorithm $y = f(x_1, \ldots, x_n)$ that transforms the inputs $x_1, \ldots, x_n$ into the result y. In mathematical terms, this dependence represents a *function*, so we need to study functions.

In line with what we discussed, we want to analyze processes which are invariant with respect to certain transformations. Thus, we need to study functions which are invariant with respect to these transformations.

In particular, since changing the measuring unit and/or starting point changes the numerical values but does not change the actual quantity, it is therefore reasonable to require that physical equations do not change if we simply change the measuring unit and/or change the starting point.

3.2 What Does It Mean for a Function to Be Invariant

Of course, to preserve the physical equations, if we change the measuring unit and/or starting point for one quantity, we may need to change the measuring units and/or starting points for other quantities as well. For example, there is a well-known relation $v = d/t$ between average velocity v and time t. If we change the measuring unit for measuring time, this formula remains valid—but only if we accordingly change the unit for velocity. For example, if we replace hours with seconds, then, to preserve this formula, we also need to change the unit for velocity from km/h to km/sec.

In general, a dependence $y = f(x_1, \ldots, x_n)$ is invariant if whatever appropriate natural transformation $(x_1, \ldots, x_n) \to (X_1, \ldots, X_n)$ we apply to the inputs, we should still have $Y = f(X_1, \ldots, X_n)$, where Y is obtained from y by a corresponding natural transformation.

© The Author(s), under exclusive license to Springer Nature Switzerland AG 2022
J. C. Urenda and V. Kreinovich, *Algebraic Approach to Data Processing*,
Studies in Big Data 115, https://doi.org/10.1007/978-3-031-16780-5_3

3.3   Example: Scale-Invariant Functions of One Variable

Let us consider the simplest case when we have a function of one variable $y = f(x)$, and we consider the simplest transformations—scaling. Which are then the invariant functions?

According to the above general definition, in this case, invariance means that for each $\lambda > 0$, if we changing all numerical values of the variable x to re-scaled values $X = \lambda \cdot x$, then the formula $y = f(x)$ should remain valid—i.e., we should have $Y = f(X)$—once we appropriately re-scale the weight function y as well, from y to $Y = \mu(\lambda) \cdot y$, for some function $\mu(\lambda)$.

In other words, if in the original units, we have $f(x) = y$, then in the new units, we will have $f(X) = Y$. Substituting the expressions for Y and X into this formula, we conclude that $f(\lambda \cdot x) = \mu(\lambda) \cdot y$ and thus,

$$f(\lambda \cdot x) = \mu(\lambda) \cdot f(x).$$

It is known that all measurable (in particular, all monotonic) solutions of this functional equation have the form $y = f(x) = A \cdot x^\alpha$ for some values A and α; see, e.g., [1].

This dependence is known as the *power law*. Thus, power laws are the only possible scale-invariant dependencies.

3.4   What If We Have Both Shift- and Scale-Invariance?

It is also natural to consider the dependence $y = f(x)$ for which, for each linear re-scaling of the x-scale, there is a corresponding linear re-scaling of the y-scale in which the dependence looks exactly the same. In other words, for every a_x and b_x there exist such values a_y and b_y that for each x and y, $y = f(x)$ implies that $\widetilde{y} = f(\widetilde{x})$, where $\widetilde{x} = a_x \cdot x + b_x$ and $\widetilde{y} = a_y \cdot y + b_y$.

It turns out that among all continuous dependencies—or, even more generally, among all the functions $f(x)$ which are, in some reasonable sense, definable—the only functions $f(x)$ satisfying this invariance property are linear functions

$$y = a \cdot x + b.$$

For linear functions, invariance is easy to prove. Indeed, suppose that $y = a \cdot x + b$. Multiplying both sides by a_x, we conclude that $a_x \cdot y = a \cdot (a_x \cdot x) + a_x \cdot b$. Here, $a_x \cdot x = \widetilde{x} - b_x$, so we get $a_x \cdot y = a_x \cdot \widetilde{x} + a_x \cdot b - a \cdot b_x$. If we add a constant $c = b - (a_x \cdot b - a \cdot b_x)$ to both sides of this equality, we conclude that $a_x \cdot y + c = a \cdot \widetilde{x} + b$, i.e., that $\widetilde{y} = a \cdot \widetilde{x} + b$, where the coefficients in the expression $\widetilde{y} = a_y \cdot y + b_y$ are equal to $a_y = a_x$ and $b_y = c$.

That only linear functions have this property is more difficult to prove; see, e.g., [2].

3.5 Which Families of Functions Are Invariant: Case of Shift-Invariance

Why families of functions. In the previous sections, we considered the case when we have a singe function $f(x)$. In practice, in different situations, we may have different functions, i.e., we have the whole *family* of functions.

How to describe a family of functions. A natural way to describe a family of functions is to select some basis $e_1(x)$, ..., $e_d(x)$, and consider the family h of all the functions of the type $d(x) = c_1 \cdot e_1(x) + \cdots + c_d \cdot e_d(x)$.

Which families are shift-invariant? Shift-invariance means that for each function $d(x)$ from the family h and for each real number a, the function $d(x + a)$ also belongs to h. In particular, this is true for the basis functions $e_1(x)$, ..., $e_d(x)$. Thus, for each i and a, there exist coefficients $c_{ij}(a)$ depending on a for which

$$e_i(x + a) = c_{i1}(a) \cdot e_1(x) + \cdots + c_{id}(a) \cdot e_d(x). \tag{3.1}$$

In particular, for each i, if we select d different values $x_1, \ldots, x_d$, then we get the following system of d linear equations for determining the coefficients $c_{ij}(a)$:

$$e_i(x_1 + a) = c_{i1}(a) \cdot e_1(x_1) + \cdots + c_{id}(a) \cdot e_d(x_1),$$

$$\cdots$$

$$e_i(x_d + a) = c_{i1}(a) \cdot e_1(x_d) + \cdots + c_{id}(a) \cdot e_d(x_d).$$

Here, the coefficients $e_j(x_k)$ are constants, so the values $c_{ij}(a)$ are linear combinations of the right-hand sides $e_i(x_k + a)$. Since the functions $e_i(x)$ are differentiable, we conclude that the values $c_{ij}(a)$ are also differentiable functions of a.

So, both sides of the equality (3.1) are differentiable. Thus, we can differentiate them with respect to a and then plug in $a = 0$. As a result, we get the following system of differential equations:

$$e_1'(x) = C_{11} \cdot e_1(x) + \cdots + C_{1d} \cdot e_d(x),$$

$$\cdots$$

$$e_d'(x) = C_{d1} \cdot e_1(x) + \cdots + C_{dd} \cdot e_d(x),$$

where $C_{ij} \stackrel{\text{def}}{=} c_{ij}'(0)$.

In other words, for the functions $e_1(x)$, ..., $e_d(x)$, we get a system of linear differential equations with constant coefficients. It is known that each solution of such system is a linear coefficient of the functions

$$x^p \cdot \exp(\alpha \cdot x), \tag{3.2}$$

where α is an eigenvalue of the matrix $\|C_{ij}\|$—which is, in general, a complex number $\alpha = a + \mathrm{i} \cdot b$, and p is a natural number which does not exceed the multiplicity of this eigenvalue. In real-number terms, we get a linear combination of the expressions $t^p \cdot \exp(a \cdot t) \cdot \sin(b \cdot t + \varphi)$.

So, every function form a shift-invariant family is a linear combination of such functions.

3.6 Which Families of Functions Are Invariant: Case of Scale-Invariance

Scale-invariance means that for each function $d(x)$ from the family h and for each positive real number $\lambda > 0$, the function $d(\lambda \cdot x)$ also belongs to h. In particular, this is true for the basis functions $e_i(x)$.

If we introduce an auxiliary variable $X \stackrel{\text{def}}{=} \ln(x)$, then replacing x with $\lambda \cdot x$ corresponds to replacing X with $X + a$, where $a \stackrel{\text{def}}{=} \ln(\lambda)$. So, for the correspondingly re-scaled functions $E_i(X) \stackrel{\text{def}}{=} e_i(\exp(X))$, we conclude that for each such function and for each real number a, the function $E_i(X + a)$ is a linear combination of functions $E_1(X)$, ..., $E_d(X)$. We already know, from Part 8 of this proof, that this implies that each function $E_i(X)$ is a linear combination of the functions $X^p \cdot \exp(\alpha \cdot X)$. Thus, each function $e_i(x) = E_i(\ln(x))$ is a linear combination of expressions

$$(\ln(x))^p \cdot \exp(\alpha \cdot \ln(x)) = (\ln(x))^p \cdot x^\alpha, \tag{3.3}$$

In real-number terms, taking into account that $\alpha = a + \mathrm{i} \cdot b$, we get a linear combination of the expressions

$$(\ln(f))^p \cdot \exp(a \cdot \ln(f)) \cdot (\cos(b \cdot \ln(f)) + \mathrm{i} \cdot \sin(b \cdot \ln(f))).$$

Here, $\exp(a \cdot \ln(f)) = (\exp(\ln(f))^a = f^a$, so the above expression has the form

$$(\ln(f))^p \cdot f^a \cdot (\cos(b \cdot \ln(f)) + \mathrm{i} \cdot \sin(b \cdot \ln(f))).$$

So, every function form a scale-invariant family is a linear combination of such functions.

3.7 What If We Have Both Shift- and Scale-Invariance

We have shown that a function from a shift-invariant family can be represented as linear combinations of the expressions (3.2), while a function from a scale-invariant family is a linear combination of expressions (3.3). One can see that the only possibility for a function to be represented both in forms (3.2) and (3.3) is to avoid logarithms and exponential functions altogether, i.e., to have $e_i(x)$ equal to a linear combination of the terms x^p for natural p, i.e., to have all functions $e_i(x)$ polynomials.

Thus, if we have a shift- and scale-invariant family, then each function from this class is a polynomial—as a linear combination of d polynomials $e_i(x)$.

3.8 Which Linear Transformations Are Shift-Invariant

In the previous sections of this chapter, we considered the case when a quantity y depends on the values of the quantities x_i—meaning that the values $y(t)$ of the quantity y at moment t is determined by the values $x_i(t)$ of the quantities x_i at this same moment of time.

In practice, sometimes, the value $y(t)$ depends also on the previous values of the quantities x_i, i.e., on the values $x_i(s)$ for moments $s < t$. This is typical, e.g., in medical and social applications: the current state of population health $y(t)$ is determined not only by the current factor such as current nutrition, pollution, etc., but also by the past events.

In many cases, this dependence is linear. A general linear dependence has the form

$$y(t) = y_0(t) + \sum_{i=1}^{n} \int c_i(t, s) \cdot x_i(s) \, ds$$

for some coefficients $c_i(t, s)$.

In many practical situation, the system does not change much during the time when measurements are performed. So if we consider similar external factors that started t_0 moments earlier, i.e., if we take $\widetilde{x}_i(t) = x_i(t + t_0)$ instead of $x_i(t)$, then the output should change accordingly, to $\widetilde{y}(t) = y(t + t_0)$. How can we describe such invariant linear transformations?

Here, on the one hand, we have

$$\widetilde{y}(t) = y_0(t) + \sum_{i=1}^{n} \int c_i(t, s) \cdot \widetilde{x}_i(s) \, ds = y_0(t) + \sum_{i=1}^{n} \int c_i(t, s) \cdot x_i(s + t_0) \, ds,$$

which, if we introduce a new variable $\widetilde{s} = s + t_0$ for which $d\widetilde{s} = ds$, leads to

$$\widetilde{y}(t) = y_0(t) + \sum_{i=1}^{n} \int c_i(t, \widetilde{s} - t_0) \cdot x_i(\widetilde{s}) \, d\widetilde{s}.$$

On the other hand,

$$\widetilde{y}(t) = y(t + t_0) = y_0(t + t_0) + \sum_{i=1}^{n} \int c_i(t + t_0, s) \cdot x_i(s) \, ds.$$

So, for all inputs $x_i(t)$, we should have

$$y_0(t) + \sum_{i=1}^{n} \int c_i(t + t_0, s) \cdot x_i(s) \, ds = y_0(t + t_0) + \sum_{i=1}^{n} \int c_i(t, s - t_0) \cdot x_i(s) \, ds.$$

Two linear functions coincide if:

- their free terms coincide, i.e., in this case, $y_0(t) = y_0(t + t_0)$ for all t and t_0, and
- the coefficients at all the unknown (in this case, $x_i(s)$) coincide, so we must have

$$c_i(t + t_0, s) = c_i(t, s - t_0).$$

From the equality $y_0(t) = y_0(t + t_0)$, we conclude that $y_0(t)$ is constant.

From the equality $c_i(t + t_0, s) = c_i(t, s - t_0)$, for every two values v_1 and v_2, we can take $s = t_0 = v_2$ and $t = v_1 - v_2$ and conclude that $c_i(v_1, v_2) = c_i(v_1 - v_2, 0)$, i.e., that $c_i(v_1, v_2) = z_i(v_1 - v_2)$, where we denoted $z_i(v) \stackrel{\text{def}}{=} c_i(v, 0)$. Substituting $y_0(t) = \text{const}$ and the above expression for $c_i(v_1, v_2)$ into the formula that describes the relation between $x_i(t)$ and $y(t)$, we conclude that

$$y(t) = y_0 + \sum_{i=1}^{n} \int z_i(t - s) \cdot x_i(s) \, ds.$$

Comment. Each term $\int z_i(t - s) \cdot x_i(s) \, ds$ is known as a *convolution* of functions $z(t)$ and $x_i(t)$.

References

1. Aczél, J., Dhombres, J.: Functional Equations in Several Variables. Cambridge University Press (2008)
2. Aczel, J., Dhombres, J.: Functional Equations in Several Variables. Cambridge University Press, Cambridge, UK (1989)

Chapter 4
What Is the General Relation Between Invariance and Optimality

4.1 What Is an Optimality Criterion

Out of all possible alternatives, we want to select an alternative which is, in some reasonable sense, optimal. For this, we need to be able to describe when some alternatives are better than others.

In practice, usually, optimality criteria are described in numerical form: we have an objective function $f(a)$ that assigns a numerical value to each possible alternative a, and we want to select an alternative for which this value is the largest possible (or, depending on the context, the smallest possible). For example, a company want to maximize its profit, a city wants to upgrade its road system so as to minimize the average travel time, etc.

However, often, we need to go somewhat beyond this approach. Indeed, for example, a company may have two (or more) different projects that lead to the same expected profit. In this case, we can use this non-uniqueness to optimize something else—e.g., out of all most-profitable projects, select the one that leads to the smallest possible long-term environmental impact. In this case, we have a more complex criterion for comparing alternatives: instead of saying that an alternative a is better than the alternative b if $f(a) > f(b)$, we say that a is better if either $f(a) > f(b)$ or $f(a) = f(b)$, and $g(a) > g(b)$ for some other numerical criterion $g(a)$. If this still does not select us a unique alternative, we can optimize yet something else, etc. In view of this possibility, in this chapter, we do not restrict ourselves to numerical optimization criteria and use the most general definition of the optimality criterion, when:

- for some pairs of alternatives a and b, we know that a is better (we will denote it by $b < a$),
- for some pairs of alternatives a and b, we know that b is better ($a < b$), and
- for some pairs of alternatives a and b, a and b are of the same value (we will denote it by $a \sim b$).

© The Author(s), under exclusive license to Springer Nature Switzerland AG 2022
J. C. Urenda and V. Kreinovich, *Algebraic Approach to Data Processing*,
Studies in Big Data 115, https://doi.org/10.1007/978-3-031-16780-5_4

Clearly, if b is better than a and c is better than b, then c should be better than a, etc. Thus, we arrive at the following definition:

Definition 4.1 Let A be a set; elements of this set will be called *alternatives*. By an *optimality criterion*, we mean a pair of binary relations $(<, \sim)$ on the set A for which the following properties hold:

- if $a < b$ and $b < c$, then $a < c$;
- if $a < b$ and $b \sim c$, then $a < c$;
- if $a \sim b$ and $b < c$, then $a < c$;
- if $a \sim b$ and $b \sim c$, then $a \sim c$;
- if $a \sim b$, then $b \sim a$;
- if $a < b$, then we cannot have $b < a$ or $a \sim b$.

Comment. Such a pair of relations is sometimes called a *partial pre-order*.

Definition 4.2 Let $(<, \sim)$ be an optimality criterion on a set A. An alternative a_{opt} is called *optimal* with respect to this criterion if for every alternative $a \in A$, we have $a < a_{\text{opt}}$ or $a \sim a_{\text{opt}}$.

4.2 We Need a Final Optimality Criterion

If an optimality criterion does not select any alternative as optimal, this means that this criterion still needs work—this may happen if for most pairs of alternatives, this criterion does not tell us which alternative is better. So, for the optimality criterion to be useful, it must select *at least one* optimal alternative.

If the criterion selects *several* alternatives as optimal, this means—as we have mentioned earlier—that this criterion is not final: we can use the resulting non-uniqueness to optimize something else, i.e., in effect, to come up with a better optimality criterion. If for this better criterion, we still have several optimal alternatives, we can (and should) modify this criterion even further, etc., until we finally get a criterion for which there is exactly one optimal alternative. Thus, we arrive at the following natural definition.

Definition 4.3 We say that an optimality criterion is *final* if there exists exactly one alternative which is optimal with respect to this criterion.

4.3   It Is Often Reasonable to Require That the Optimality Criterion Be Invariant

In the situations, when there are natural transformations T that do not change the situation—e.g., changing the measuring unit—it makes sense to require expect that which alternative is better should not change after this transformation.

Definition 4.4 Let A be a set of alternatives, let $(<, \sim)$ be an optimality criterion of the set A, and let $\mathscr{T}$ be a class of transformations $A \to A$. We say that the optimality criterion $(<, \sim)$ is $\mathscr{T}$-*invariant* if for every $T \in \mathscr{T}$ and for all $a, b \in A$, the following two properties hold:

- if $a < b$ then $T(a) < T(b)$, and
- If $a \sim b$, then $T(a) \sim T(b)$.

4.4   Main Result of This Chapter

Proposition 4.1 *Let Let A be a set of alternatives, and let $(<, \sim)$ be a final $\mathscr{T}$-invariant optimality criterion on the set A. Then the optimal alternative a_{opt} is also $\mathscr{T}$-invariant, i.e., $T(a_{opt}) = a_{opt}$ for all $T \in \mathscr{T}$.*

Proof Since the optimality criterion is final, there exists exactly one alternative a_{opt} which is optimal with respect to this criterion, i.e., for which $a < a_{opt}$ or $a \sim a_{opt}$ for all other alternatives a.

Due to optimality, for each alternative a and for each transformation $T \in \mathscr{T}$, for the alternative $T^{-1}(a)$, we have either $T^{-1}(a) < a_{opt}$ or $T^{-1}(a) \sim a_{opt}$.

Since the optimality criterion is $\mathscr{T}$-invariant, we thus have either $T(T^{-1}(a)) < T(a_{opt})$ or $T(T^{-1}(a)) \sim T(a_{opt})$. Here, by the definition of the inverse transformation, $T(T^{-1}(a)) = a$, so we conclude that for every alternative a, we have either $a < T(a_{opt})$ or $a \sim T(a_{opt})$. By definition of optimality, this means that the alternative $T(a_{opt})$ is optimal. However, our optimality criterion is final, which means that there is only one optimal alternative. Thus, indeed, $T(a_{opt}) = a_{opt}$.

Chapter 5
General Application: Dynamical Systems

It can be proven that linear dynamical systems exhibit either stable behavior, or unstable behavior, or oscillatory behavior, or transitional behavior. Interesting, the same classification often applies to nonlinear dynamical systems as well. In this paper, we provide a possible explanation for this phenomenon, i.e., we explain why a classification based on linear approximation to dynamical systems often works well in nonlinear cases.

Results from this chapter first appeared in [1].

5.1 Problem: Why a Linear-Based Classification Often Works in Nonlinear Cases

Dynamical systems are ubiquitous. To describe the state of a real-life system at any given moment of time, we need to know the values $x = (x_1, \ldots, x_n)$ of all the quantities that characterize this system. For example, to describe the state of a mechanical system consisting of several pointwise objects, we need to know the position and velocities of all these objects. To describe the state of an electric circuit, we need to know the currents and voltages, etc.

In many real-life situation, the corresponding systems are deterministic—in the sense that the future states of the system are uniquely determined by its current state. Sometimes, to make the system deterministic, we need to enlarge its description so that it incorporates all the objects that affect its dynamics. For example, the system consisting of Earth and Moon is not deterministic in its original form—since the Sun affects its dynamics, but once we add the Sun, we get a system with a deterministic behavior.

The fact that the future dynamics of the system is uniquely determined by its current state means, in particular, that the rate $\dot{x}$ with which the system changes is

© The Author(s), under exclusive license to Springer Nature Switzerland AG 2022
J. C. Urenda and V. Kreinovich, *Algebraic Approach to Data Processing*,
Studies in Big Data 115, https://doi.org/10.1007/978-3-031-16780-5_5

also uniquely determined by its current state, i.e., that we have $\dot{x} = f(x)$, for some function $f(x)$. This equation can be described coordinate-wise, as

$$\dot{x}_i = f_i(x_1, \ldots, x_n). \tag{5.1}$$

Systems that satisfy such equations are known as *dynamical systems*; see, e.g., [2].

Simplest case: linear systems. The simplest case is when the rate of change $f_i(x_1, \ldots, x_n)$ of each variables is a linear function, i.e., when

$$\dot{x}_i = a_{i0} + \sum_{j=1}^{n} a_{ij} \cdot x_j. \tag{5.2}$$

In almost all such cases—namely, in all the cases when the matrix a_{ij} is non-degenerate—we can select constants s_i so that for the correspondingly shifted variables $y_i = x_i + s_i$, the system gets an even simpler form

$$\dot{y}_i = \sum_{j=1}^{n} a_{ij} \cdot y_j. \tag{5.3}$$

Indeed, substituting $x_i = y_i - s_i$ into the formula (5.2), and taking into account that $\dot{y}_i = \dot{x}_i$, we conclude that

$$\dot{x}_i = a_{i0} + \sum_{j=1}^{n} a_{ij} \cdot (y_j - s_j) = a_{i0} + \sum_{j=1}^{n} a_{ij} \cdot y_j - \sum_{j=1}^{n} a_{ij} \cdot s_j.$$

Thus, if we select the value s_j for which $a_{i0} = \sum_{j=1}^{n} a_{ij} \cdot s_j$ for each i, we will indeed get the formula (5.2).

For the Eq. (5.2), the general solution is well known: it is a linear combination of expressions of the type $t^k \cdot \exp(\lambda \cdot t)$, where λ is an eigenvalue of the matrix $\|a_{ij}\|$—which is, in general, a complex number $\lambda = a + i \cdot b$, and k is a natural number which does not exceed the multiplicity of this eigenvalue. In real-number terms, we get a linear combination of the expressions $t^k \cdot \exp(a \cdot t) \cdot \sin(b \cdot t + \varphi)$.

Depending on the values of λ, we have the following types of behavior:

- when $a < 0$ for all the eigenvalues, then the system is *stable*: no matter what state we start with, it asymptotically tends to the state $y_1 = \ldots = y_n = 0$;
- when $a > 0$ for at least one eigenvalue, then the system is *unstable*: the deviation from the 0 state exponentially grows with time;
- when $a = 0$ and $b \neq 0$, we get an *oscillatory behavior*; and
- when $a = b = 0$, we get a *transitional behavior*, when a system linearly (or quadratically etc.) moves from one state to another.

A similar classification works well in non-linear cases, but why? Interestingly, a similar classification works well for nonlinear dynamical systems as well, but why? In this chapter, we will try to explain this fact.

5.2 Our Explanation

We need finite-dimensional approximations. We want to describe how the state $x(t) = (x_1(t), \ldots, x_n(t))$ of a dynamical system changes with time t. In general, the set of all possible smooth functions $x_i(t)$ is infinite-dimensional, i.e., we need infinitely many parameters to describe it. However, in practice, at any given moment, we can only have finitely many parameters. Thus, it is reasonable to look for finite-parametric approximations. A natural idea is to fix some smooth functions $e_k(t) = (e_{k1}(t), \ldots, e_{kn}(t))$, $1 \leq k \leq K$, and consider linear combinations

$$x(t) = \sum_{k=1}^{K} c_k \cdot e_k(t). \tag{5.4}$$

Shift-invariance. For dynamical systems, there is no fixed moment of time. The equations remain the same if we change the starting point for measuring time, i.e., if we replace the original temporal variable t with the new variable $t' = t + t_0$.

It is therefore reasonable to require that the approximating family (5.4) be invariant with respect to the same transformation, i.e., in other words, that all shifted functions $e_k(t + t_0)$ can also be represented in the same form (5.4). As we have shown in Chap. 3, this implies that the functions $e_{ki}(t)$ satisfy the system of linear differential equations with constant coefficients—and we have already mentioned that the solutions to such systems are exactly the functions leading to a known classification of linear dynamical system behaviors.

This explains why for nonlinear systems, we also naturally observe similar types of behavior.

References

1. Urenda, J., Kreinovich, V.: Why a classification based on linear approximation to dynamical systems often works well in nonlinear cases. In: Proceedings of the Annual Conference of the North American Fuzzy Information Processing Society NAFIPS'2020, Redmond, Washington, 20–22 Aug. 2020
2. Hirsch, M.W., Smale, S., Devaney, R.L.: Differential Equations, Dynamical Systems, and an Introduction to Chaos. Academic Press, Waltham, Massachisetts (2013)

Chapter 6
First Application to Physics: Why Liquids?

6.1 Two Applications to Physics: Summary

In this chapter and in the following chapter, we describe applications of algebraic approach to physics. While physics has had many successes, it also has many remaining challenges—this can be seen from the fact that every year, new spectacular results get Nobel Prizes. However, most of these challenges require a large amount of preliminary knowledge to explain. To illustrate the efficiency of the algebraic approach, we decided to select two problems from two different areas of physics for which the corresponding challenge is easy to understand.

The first challenge may be surprising to most readers: how to explain why liquids exist. In 1977, the renowned physicist Victor Weisskopf challenged the physics community to provide a fundamental explanation for the existence of the liquid phase of matter. A recent essay confirms that Weisskopf's 1977 question remains a challenge. In this chapter, we use natural symmetry ideas to show that liquids are actually a natural state between solids and gases.

Main results from this chapter first appeared in [1].

The following chapter deals with another easy-to-understand challenge—how to explain the observed warping of our Galaxy.

6.2 Problem: Why Liquids?

Weisskopf's challenge: original formulation. In his 1977 essay [2], the renowned physicist Victor Weisskopf challenged the physics community to provide a fundamental explanation for the existence of liquids. Solid bodies, in their crystal form, with their natural symmetries, correspond to the state with the smallest possible energy—the state that most materials take at the absolute zero temperature ($0°$ K, which is approximately $-273°$C). Gases, with no restriction on their shapes, are also

© The Author(s), under exclusive license to Springer Nature Switzerland AG 2022
J. C. Urenda and V. Kreinovich, *Algebraic Approach to Data Processing*,
Studies in Big Data 115, https://doi.org/10.1007/978-3-031-16780-5_6

a natural state—they correspond to high temperatures. When heated, solid bodies eventually become gases—usually, with an intermediate liquid state. But how to explain this intermediate state?

Weisskopf's challenge: current status. A recent essay [3], while overviewing the progress in answering this question, confirms that Weisskopf's 1977 question remains a challenge.

What we do in this chapter. In this chapter, we use natural symmetry ideas to show that liquids are actually a natural state between solids and gases.

6.3 Towards a Formulation of the Problem in Precise Terms

Local transformations. In nature, most processes are smooth, so it makes sense to consider only smooth (differentiable) transformations.

All states allow shifts, so we will look for transformations beyond shifts.

Locally, in the vicinity of each point $(x_1^{(0)}, \ldots, x_n^{(0)})$, i.e., for points

$$(x_1, \ldots, x_n) = (x_1^{(0)} + \Delta x_1, \ldots, x_n^{(0)} + \Delta x_n),$$

we can expand each smooth transformation

$$(x_1, \ldots, x_n) \to (y_1, \ldots, y_n) = (f_1(x_1, \ldots, x_n), \ldots, f_n(x_1, \ldots, x_n))$$

in Taylor series and keep only linear terms in this expansion. Then:

$$y_i = f_i(x_1, \ldots, x_n) = f_i(x_1^{(0)} + \Delta x_1, \ldots, x_n^{(0)} + \Delta x_n) \approx y_i^{(0)} + \sum_{j=1}^{n} a_{ij} \cdot \Delta x_j,$$

where $y_i^{(0)} \stackrel{\text{def}}{=} (x_1^{(0)}, \ldots, x_n^{(0)})$ and $a_{ij} \stackrel{\text{def}}{=} \dfrac{\partial f_i}{\partial x_j}_{|(x_1^{(0)}, \ldots, x_n^{(0)})}$. Since shift is always possible, we can apply the shift by $x_i^{(0)} - y_i^{(0)}$, thus getting a new transformation

$$(x_1^{(0)} + \Delta x_1, \ldots, x_n^{(0)} + \Delta x_n) \to (y_1, \ldots, y_n) =$$

$$\left(x_1^{(0)} + \sum_{j=1}^{n} a_{1j} \cdot \Delta x_j, \ldots, x_n^{(0)} + \sum_{j=1}^{n} a_{nj} \cdot \Delta x_j \right).$$

In terms of the differences $\Delta y_i \stackrel{\text{def}}{=} y_i - x_i^{(0)}$, this local transformation becomes linear homogeneous:

$$\Delta y_i = \sum_{j=1}^{n} a_{ij} \cdot \Delta x_j.$$

So, locally, it is sufficient to consider linear transformations.

Local Lie group: a brief reminder. Clearly, if two transformations are possible, then their composition is also possible. Also, if a transformation is possible, then an inverse transformation is also possible. In mathematics, classes of transformations which are closed under composition and under taking the inverse are known as *groups*. In these terms, transformations form a group.

The set of all possible transformations usually smoothly depends on some parameters. In mathematics, such transformation groups are called *Lie groups*. Local transformations form a *local Lie group*.

Local (linear) transformations are uniquely determined by the corresponding matrices a_{ij}. One can easily see that the composition of two transformations corresponds to the product of the two matrices, and the inverse transformation corresponds to an inverse matrix. Thus, the local Lie group is a class of matrices which is closed under matrix multiplication and taking the inverse.

Lie algebras: a brief reminder. Each transformation occurs during a certain time period T, during which the original state smoothly turns into the new state. At each moment of time $t \leq T$, we have some intermediate stage of this transformation. We can thus divide the interval of width T into several small subintervals of size $\Delta t \ll T$, and consider the whole transformation as a composition of transformations from moment 0 to moment Δt, from moment Δt to moment $2\Delta t$, etc., until we reach moment T. Let us denote the transformation from moment t to moment t' by $a_{ij}(t, t')$. In these terms, the transformation from a moment t to the next moment $t + \Delta t$ has the form $a_{ij}(t, t + \Delta t)$. For $t' = t$, the transformation is the identity matrix $a_{ij}(t, t) = \Delta_{ij}$, where $\Delta_{ii} = 1$ for all i and $\Delta_{ij} = 0$ for all $i \neq j$.

Since the transformation is smooth, we can expand the dependence

$$a_{ij}(t, t + \Delta t)$$

in Taylor series and keep only terms linear in Δt in this expansion:

$$a_{ij}(t, t + \Delta t) = a_{ij}(t, t) + b_{ij}(t) \cdot \Delta t = \delta_{ij} + b_{ij}(t) \cdot \Delta t, \quad (6.1)$$

where $b_{ij}(t) \stackrel{\text{def}}{=} \dfrac{\partial a_{ij}(t, t')}{\partial t'}\Big|_{t'=t}$.

In mathematical terms, the above transformation (6.1)—corresponding to very small time intervals Δt—is called *infinitesimal*, and the class of all such transformations is known as a *Lie algebra*.

Each transformation can be represented as a composition of transformations corresponds to small time intervals Δt. Once we know the corresponding Lie algebra, we can describe each such transformation with good accuracy and thus, describe the original transformation as their composition. The smaller Δt, the more accurate this representation. This means that, once we know the Lie algebra, we can determine all possible transformations with any desired accuracy.

So, to describe the class of all possible transformations, it is sufficient to describe the corresponding Lie algebra.

Natural properties of Lie algebras: reminder. If we have two infinitesimal transformations, with matrices $a'_{ij} = \delta_{ij} + b'_{ij} \cdot \Delta t$ and $a''_{ij} = \delta_{ij} + b''_{ij} \cdot \Delta t$, then the matrix a_{ij} describing their composition is equal to the product of these matrices:

$$a_{ik} = \sum_{j=1}^{n} a'_{ij} \cdot a''_{jk} = \sum_{j=1}^{n} (\delta_{ij} + b'_{ij} \cdot \Delta t) \cdot (\delta_{jk} + b''_{jk} \cdot \Delta t).$$

Opening parentheses and ignoring terms proportional to $(\Delta t)^2$, we conclude that

$$a_{ik} = \sum_{j=1}^{n} \delta_{ij} \cdot \delta_{jk} + \sum_{i=1}^{n} \delta_{ij} \cdot b''_{jk} \cdot \Delta t + \sum_{j=1}^{n} b'_{ij} \cdot \delta_{jk} \cdot \Delta t.$$

Taking into account that only the diagonal values of the unit matrix δ_{ij} are non-zeros—and that these diagonal elements are equal to 1—we conclude that

$$a_{ik} = \delta_{ik} + (b'_{ik} + b''_{ik}) \cdot \Delta t.$$

Thus, the composition of two transformations corresponds to the sum of the matrices from Lie algebra. So, the Lie algebra should be closed under addition.

Similarly, one can show that the inverse corresponds to $-b_{ij}$, and that, in general, for each real number λ and for each matrix b_{ij}, the Lie algebra also contains the matrix $\lambda \cdot b_{ij}$. Combining this property with addition, we can conclude that each Lie algebra contains an arbitrary linear combination of its elements.

Resulting formulation of the problem. For a solid body, the only possible transformations are shifts and rotations. Thus, the only possible local transformations are rotations. It is known that the corresponding Lie algebra A_s (s for solid) consists of all antisymmetric matrices b_{ij}, i.e., matrices for which $b_{ji} = -b_{ij}$ for all i and j.

For the gas, all smooth transformations are possible. Thus, the corresponding Lie algebra L_g (g for gas) consists of all the matrices b_{ij}.

In these terms, the question is: what are the Lie algebras L which are strictly larger than L_s but strictly smaller than L_g, i.e., algebras for which $L_s \subset L \subset L_g$?

6.4 Main Result of This Chapter

Proposition 6.1 *There are exactly two Lie algebras L that strictly contain the algebra L_s of all antisymmetric matrices and that are strictly contained in the algebra L_g of all matrices:*

- *the algebra L_l that consists of all matrices with zero trace* $\mathrm{Tr}(b) \overset{\mathrm{def}}{=} \sum_{i=1}^{3} b_{ii} = 0$, *and*

- *the algebra L_a obtained by adding all scalar matrices $\lambda \cdot \delta_{ij}$ to antisymmetric ones.*

Discussion. Liquids are (largely) incompressible—this is their main difference from gases. This means that they are characterized by transformations that preserve volume. In terms of Lie algebras, preserving volume means exactly $\mathrm{Tr}(b) = 0$. Thus, our result indeed explains the existence of liquids.

We also have another case. These additional elements of Lie algebra correspond to increasing and decreasing the size of the object without changing its proportions. This may correspond to some yet unknown state of nature.

Proof $1°$. Let L be a Lie algebra that strictly contains L_s and that is strictly contained in L_g. Let us first show that the algebra L contains a matrix b_{ij} if and only if it contains its symmetric part $b_{ij}^{\mathrm{sym}} \overset{\mathrm{def}}{=} \dfrac{b_{ij} + b_{ji}}{2}$.

Indeed, each matrix b_{ij} can be represented as a sum of its symmetric part and an antisymmetric part $b_{ij}^{\mathrm{asym}} \overset{\mathrm{def}}{=} \dfrac{b_{ij} - b_{ji}}{2}$. Since the class L is closed under addition and contains the class L_g of all antisymmetric matrices, this means that if the symmetric part of b_{ij} is in L, then the original matrix b_{ij} is also in L, as the sum of two matrices b_{ij}^{sym} and b_{ij}^{asym} from the class L.

Vice versa, since the class L is closed under linear combination, for each matrix b_{ij}, the class L contains the matrix $b_{ij} - b_{ij}^{\mathrm{asym}}$, which is exactly the symmetric part of the matrix b_{ij}.

$2°$. Due to Part 1 of the proof, to describe all the matrices from the class L, it is sufficient to describe all symmetric matrices from this class.

$3°$. Since the Lie algebra contains all antisymmetric matrices, the corresponding transformation group contains all rotations T. One can easily check that if we first apply the rotation T, then an infinitesimal transformation with matrix $B = \|b_{ij}\|$, and then inverse rotation T^{-1}, we get an infinitesimal transformation with matrix $T^{-1}BT$. Thus, with each matrix B, the Lie algebra L contains all the matrices obtained from it by a rotation.

It is known that each symmetric matrix, by an appropriate rotation—in which coordinates axes are rotated into the matrix's eigenvectors—can be transformed into a diagonal form. Thus, to describe all symmetric matrices from L, it is sufficient to describe all diagonal matrices from the class L.

$4°$. Let us prove that if L contains at least one matrix with $\mathrm{Tr}(b) \neq 0$, then it contains all scalar matrices.

Indeed, if the algebra A contains one matrix with non-zero trace, then its diagonal form $b = \mathrm{diag}(b_{11}, b_{22}, b_{33})$ will have the same trace $\mathrm{Tr}(b) = b_{11} + b_{22} + b_{33} \neq 0$. Thus, due to rotation-invariance, it also contains matrices $\mathrm{diag}(b_{22}, b_{33}, b_{11})$ and $\mathrm{diag}(b_{33}, b_{11}, b_{22})$ obtained from the original one by a rotation that changes the coordinate axes. Hence, it contains the sum of these three matrices—which is a scalar matrix $\mathrm{Tr}(b) \cdot \delta_{ij}$. Every other scalar matrix can be obtained from this one by multiplying by a number; thus, every scalar matrix indeed belongs to L.

$5°$. Let us now prove that if L contains at least one non-scalar symmetric matrix, then it contains all matrices with $\mathrm{Tr}(b) = 0$.

Indeed, let us assume that L contains a non-scalar symmetric matrix. Then, its diagonal form $b = \mathrm{diag}(b_{11}, b_{22}, b_{33})$ is also non-scalar and symmetric. If $\mathrm{Tr}(b) \neq 0$, then, by Part 4 of this proof, all scalar matrices are also in L, in particular, a scalar matrix $(\mathrm{Tr}(b)/3) \cdot \delta_{ij}$ with the same trace. Subtracting this scalar matrix from b, we get a new non-scalar symmetric matrix from the algebra L whose trace is 0. Thus, without losing generality, we can assume that the original non-scalar symmetric matrix has trace 0. Since the matrix is non-scalar, at least two diagonal values b_{ii} are different from each other. Without losing generality, we can assume that $b_{11} \neq b_{22}$.

If we rotate by $90°$ in the x_1-x_2 plane, we get a new diagonal matrix $b'_{ij} = \mathrm{diag}(b_{22}, b_{11}, b_{33})$. Subtracting b'_{ij} from b_{ij}, we get yet another diagonal matrix from the algebra L: the matrix $\mathrm{diag}(b_{11} - b_{22}, b_{22} - b_{11}, 0)$. Since the Lie algebra is closed under multiplication by a scalar, we thus conclude that the diagonal matrix $\mathrm{diag}(1, -1, 0)$ also belongs to L.

By rotation, we can conclude that $\mathrm{diag}(0, 1, -1) \in L$. Now, each diagonal matrix with 0 trace has the form $\mathrm{diag}(p, q, -(p + q))$. We can represent this matrix as a linear combination of matrices from L:

$$\mathrm{diag}(p, q, -(p + q)) = p \cdot \mathrm{diag}(1, -1, 0) + (p + q) \cdot \mathrm{diag}(0, 1, -1).$$

Thus, every diagonal matrix with 0 trace belongs to L, and hence, every symmetric matric with 0 trace belongs to L.

$6°$. Now, we are ready to prove the proposition. Since $L_s \subset L$ and $L \neq L_s$, we conclude that L must contain at least one matrix which is not antisymmetric, and thus, must contain at least one symmetric matrix—namely, the result of its symmetrization.

If L contains a non-scalar matrix and a matrix with non-zero trace, then L contains all scalar matrices and all symmetric matrices with zero trace. Each symmetric matrix can be represented as a sum of the corresponding scalar part and zero-trace part, so this would mean that L contains all symmetric matrices—and thus, all of them. This contradicts to our assumption that $L \neq L_g$. So, L cannot contain both types of matrices.

Thus, we have two options:

- The first option is that L contains a non-scalar matrix. In this case, according to Part 5 of the proof, L contains all matrices with 0 trace, and, as we have just proved, it cannot contain any matrix with non-zero trace. In this case, L is the set of all matrices with 0 trace, i.e., $L = L_l$.
- The second option is that L contains a symmetric matrix with non-zero trace. In this case, L contains all scalar matrices—and, as we have just shown, it cannot contain any non-scalar symmetric matrix. This is the second case $L = L_a$.

The proposition is proven.

References

1. Urenda, J.C., Kreinovich, V.: Why liquids? a symmetry-based solution to Weisskopf's challenge. J. Uncertain Syst. **13**(3), 224–228 (2019)
2. Weisskopf, V.F.: About liquids. Trans. New York Acad. Sci. **38**, 202–218 (1977)
3. Evans, R., Frenkel, D., Dijkstra, M.: From simple liquids to colloids and soft matter. Phys. Today **38**(2), 38–39 (2019)

Chapter 7
Second Application to Physics: Warping of Our Galaxy

In the first approximation, the shape of our Galaxy—as well as the shape of many other celestial bodies—can be naturally explained by geometric symmetries and the corresponding invariances. As a result, we get the familiar shape of a planar spiral. A recent more detailed analysis of our Galaxy's shape has shown that the Galaxy somewhat deviates from this ideal shape: namely, it is not perfectly planar, it is somewhat warped in the third dimension. In this chapter, we show that the empirical formula for this warping can also be explained by geometric symmetries and invariance.

Main results from this chapter first appeared in [1].

7.1 Formulation of the Problem

In the first approximation, geometry explains shapes of celestial bodies such as galaxies. In modern physics, symmetries—including geometric symmetries—and related invariances play a very important role; see, e.g. [2, 3].

As part of this general trend, it is known that the geometric shapes of most celestial bodies can be explained by symmetry groups and corresponding invariances; see, e.g., [4–6]. Namely, in the beginning, the Universe was homogeneous, isotropic, and scale-invariant—i.e., in geometric terms, it was invariant with respect to shifts, rotations, and homotheties. Such highly symmetric matter distributions are, however, unstable, so spontaneous symmetry breaking leads us to states with fewer invariances. In general, the more invariances are preserved, the more probable the corresponding transition. Thus, the most probable transition from the original fully symmetric state is to a planar shape ("pancake", a typical shape of a proto-galaxy), and the next most probably transition is to a planar logarithmic spiral—a generic planar shape with a

© The Author(s), under exclusive license to Springer Nature Switzerland AG 2022
J. C. Urenda and V. Kreinovich, *Algebraic Approach to Data Processing*,
Studies in Big Data 115, https://doi.org/10.1007/978-3-031-16780-5_7

1-dimensional symmetry group. In perfect accordance with this symmetry analysis, the shape of our Galaxy is indeed well approximated by a planar logarithmic spiral.

Comment. It should be mentioned that eventually, each celestial body gets transformed into the most stable state—of a sphere or, in case of rotation, of a rotational ellipsoid [4–6].

A recent empirical formula for a more detailed description of our Galaxy's shape. A recent more detailed analysis of the Galaxy's shape [7] has shown that this shape somewhat deviates from the above symmetric form. Specifically, the spiral shape is still there, but the Galaxy is not exactly planar. In the corresponding cylindrical coordinates (r, φ, z) in which the plane has the form $z = 0$, the z-coordinate of the actual Galaxy is only equal to 0 up to a certain threshold distance r_d from the center ($r \leq r_d$). For the distances $r > r_d$, the following empirical formula describes the actual shape:

$$z = z_0 \cdot (r - r_d)^2 \cdot \sin(\varphi - \varphi_0), \tag{7.1}$$

for some parameters z_0 and φ_0.

What we do in this chapter. In this chapter, we show that this empirical formula can also be naturally explained by the corresponding geometric symmetries and invariances.

7.2 Analysis of the Problem and the Resulting Explanation

Main idea. Instead of the plane $z = 0$, we have a warped shape. To be more precise, we have a planar shape until $r - r_d$, and then the shape $z(r, \varphi)$ becomes warped.
   The original planar shape $z = 0$ is invariant with respect to all rotations

$$\varphi \to \varphi + \alpha$$

around the Galaxy's center. The observed shape (7.1) is, however, not rotation-invariant. Thus, the corresponding function $z(r, \varphi)$ is not rotation invariant. Since we cannot have a single rotation-invariant function, a natural idea is to assume that the actual shape-describing function $z(r, \varphi)$ belongs to a few-parametric rotation-invariant *family* of functions

$$C_1 \cdot z_1(r, \varphi) + \cdots + C_k \cdot z_k(r, \varphi),$$

where $z_1(r, \varphi), \ldots, z_k(r, \varphi)$ are fixed functions, and $C_1, \ldots, C_k$ are arbitrary coefficients.
   Let us select a family with as fewer parameters as possible—since in general, the fewer parameters we need to describe a physical phenomenon, the more convincing

our explanation [2]: with a large number of parameters, we can explain anything by properly adjusting the values of these parameters.

Simplest case $k = 1$ does not help. The simplest case if $k = 1$, when we have a family $\{C_1 \cdot z_1(r, \varphi)\}$. For this family, invariance means that if we rotate by an angle α—i.e., if we replace φ with $\varphi + \alpha$—then we still get a function from the same family, i.e., $z_1(r, \varphi + \alpha) = C(\alpha) \cdot z_1(r, \varphi)$. For each r, we have a functional equation whose solutions are known (see, e.g., [8]): $z_1(r, \varphi) = z_1(r, 0) \cdot \exp(a(r) \cdot \varphi)$ for some $a(r)$. Since rotation by $\alpha = 2\pi$ does not change anything, we should have $z_1(r, \varphi + 2\pi) = z_1(r, \varphi)$. This implies that $a(r) = 0$ and thus, that the corresponding function $z_1(r, \varphi)$ does not depend on the angle φ at all—and we know that it depends.

Thus, to adequately describe the actual shape of our Galaxy, we need to consider families with more parameters—at least $k = 2$.

Case of $k = 2$. In this case, we have a family $\{C_1 \cdot z_1(r, \varphi) + C_2 \cdot z_2(r, \varphi)\}$. The fact that this family in rotation-invariant means that the rotated functions $z_1(r, \varphi + \alpha)$ and $z_2(r, \varphi + \alpha)$ should also belong to this family, i.e., that we should have $z_1(r, \varphi + \alpha) = C_{11}(\alpha) \cdot z_1(r, \varphi) + C_{12}(\alpha) \cdot z_2(r, \varphi)$ and $z_2(r, \varphi + \alpha) = C_{21}(\alpha) \cdot z_1(r, \varphi) + C_{22}(\alpha) \cdot z_2(r, \varphi)$.

From the physical viewpoint, it makes sense to assume that the shapes are smooth and that, that the function $z(r, \varphi)$ is differentiable. Thus, it makes sense to restrict ourselves to the case when both functions $z_i(r, \varphi)$ are differentiable. By using Cramer's formulas, we can explicitly express all four values $C_{ij}(\alpha)$ as rational expressions in terms of the values of the functions $z_i(r, \varphi)$; thus, the corresponding functions $C_{ij}(\alpha)$ are also differentiable. Differentiating the above equations with respect to α and taking $\alpha = 0$, we get, for each r, the following system of linear differential equations with constant coefficients:

$$z_i'(r, \varphi) = c_{i1} \cdot z_1(r, \varphi) + c_{i2} \cdot z_2(r, \varphi) \ (i = 1, 2),$$

where z' denotes derivative with respect to φ and $c_{ij} \stackrel{\text{def}}{=} C_{ij}'(0)$.

It is known that a general solution to such a system is a linear combination of functions $\varphi^m \cdot \exp(a \cdot \varphi) \cdot \sin(b \cdot \varphi + \varphi_0)$. The requirement that $z_i(r, \varphi + 2\pi) = z_i(r, \varphi)$ eliminates powers and exponents and leaves only sines and cosines with integer b. In line with the general argument about minimizing the number of parameters, let us restrict ourselves to the simplest case $b = 1$. Then, for each i, we have $z_i(r, \varphi) = A_i(r) \cdot \sin(\varphi) + B_i(r) \cdot \cos(\varphi)$ for some $A_i(r)$ and $B_i(r)$. The actual function $z_1(r, \varphi)$ is a linear combination of these functions, so we have $z(r, \varphi) = A(r) \cdot \sin(\varphi) + B(r) \cdot \cos(\varphi)$ for appropriate $A(r)$ and $B(r)$.

Which functions $A(r)$ and $B(r)$ should we choose? We know that $z(r, \varphi) = 0$ for $r \leq r_d$ and, in general, $z(r, \varphi) \neq 0$ for $r > r_d$. We have assumed that the function is differentiable.

The difference $z(r, \varphi)$ from planarity is relatively small, so for $r > r_d$, it makes sense to expand the dependence $z(r, \varphi)$ in Taylor series and keep only the first few terms in this expansion. We cannot keep only linear terms—otherwise, the resulting piece-wise linear dependence $z(r, \varphi)$ on r will not be differentiable at $r = r_d$. Thus, we need to also keep quadratic terms. One can easily check that the only quadratic functions that smoothly transition to 0 at $r = r_d$ are functions $c \cdot (r - r_d)^2$ for some constant c. Both functions $A(r)$ and $B(r)$ should have this form, for appropriate coefficients c_a and c_b—since they represent $z_1(r, \varphi)$ when $\varphi = \pi/2$ and when $\varphi = 0$. Thus, for $r > r_d$, the above expression for $z(r, \varphi)$ has the form

$$z(r, \varphi) = c_a \cdot (r - r_d)^2 \cdot \sin(\varphi) + c_b \cdot (r - r_d)^2 \cdot \cos(\varphi) =$$
$$(r - r_d)^2 \cdot (c_a \cdot \sin(\varphi) + c_b \cdot \cos(\varphi)). \qquad (7.2)$$

So, we get the desired geometric explanation. The expression $c_a \cdot \sin(\varphi) + c_b \cdot \cos(\varphi)$ can be easily transformed into the form $z_1 \cdot \sin(\varphi - \varphi_0)$, with $z_0 = \sqrt{c_a^2 + c_b^2}$ and appropriate φ_0.

Thus, from the formula (7.2), we indeed get the desired expression (7.1).

Remaining open problems. In general, for celestial bodies, geometric invariances do not just explain possible shapes; they also explain relative frequencies of different shapes, prevalent directions of rotation and of the body's magnetic field, etc. [4–6].

It would therefore be nice to similarly extend the above geometric explanation of our Galaxy's shape to an explanation of other related empirical formulas—starting with other formulas presented in [7].

References

1. Urenda, J., Kosheleva, O., Kreinovich, V.: Geometric explanation for an empirical formula describing our galaxy's warping. In: University of Texas at El Paso, Department of Computer Science, Technical Report UTEP-CS-19-84 (2019)
2. Feynman, R., Leighton, R., Sands, M.: The Feynman Lectures on Physics. Addison Wesley, Boston, Massachusetts (2005)
3. Thorne, K.S., Blandford, R.D.: Modern Classical Physics: Optics, Fluids, Plasmas, Elasticity, Relativity, and Statistical Physics. Princeton University Press, Princeton, New Jersey (2017)
4. Finkelstein, A., Kosheleva, O., Kreinovich, V.: Astrogeometry: towards mathematical foundations. Int. J. Theoret. Phys. **36**(4), 1009–1020 (1997)
5. Finkelstein, A., Kosheleva, O., Kreinovich, V.: Astrogeometry: geometry explains shapes of celestial bodies. Geombinatorics V **I**(4), 125–139 (1997)
6. Li, S., Ogura, Y., Kreinovich, V.: Limit Theorems and Applications of Set Valued and Fuzzy Valued Random Variables. Kluwer Academic Publishers, Dordrecht (2002)
7. Skowron, D.M., Skowron, L., Mróz, P., Udalski, A., Pietrukowicz, P., Soszyński, I., Szymański, M.K., Poleski, R., Kozlowski, S., Ulaczyk, K., Kybicki, K., Iwanek, P.: A three-dimensional map of the Milky Way using classical Cepheid variable stars. Science **365**(6452), 478–482 (2019)
8. Aczel, J., Dhombres, J.: Functional Equations in Several Variables. Cambridge University Press, Cambridge, UK (1989)

Chapter 8
Application to Electrical Engineering: Class-D Audio Amplifiers

8.1 Applications to Engineering: Summary

In this chapter and in the following chapter, we describe applications of algebraic methods to different engineering challenges. Some engineering problems are well understood and well-formalized, so solving these problems is just a matter of mathematics and computations. However, there are many other important engineering problems for which formalization is not easy:

- some of these problems deal with user perception which is not easy to formally describe: how do we describe the quality of sound? the smoothness of a ride?
- other problems deal with processing raw materials like wood for building or soil for pavements, materials whose properties are not easy to describe in precise terms.

In these two chapters, we show that algebraic approach can be used to solve both types of challenges.

In this chapter, we deal with the fact that most current high-quality electronic audio systems use class-D audio amplifiers (D-amps, for short), in which a signal is represented by a sequence of pulses of fixed height, pulses whose duration at any given moment of time linearly depends on the amplitude of the input signal at this moment of time. In this abstract, we explain the efficiency of this signal representation by showing that this representation is the least vulnerable to additive noise (that affect measuring the signal itself) and to measurement errors corresponding to measuring time.

Results from this chapter first appeared in [1].

The next chapters deal with properties of wood.

J. C. Urenda and V. Kreinovich, *Algebraic Approach to Data Processing*, Studies in Big Data 115, https://doi.org/10.1007/978-3-031-16780-5_8

8.2   Problem: Why Class-D Audio Amplifiers Work Well?

Most current electronic audio systems use class-D audio amplifiers, where a signal $x(t)$ is represented by a sequence $s(t)$ of pulses whose height is fixed and whose duration at time t linearly depends on the amplitude $x(t)$ of the input signal at this moment of time; see, e.g., [2] and references therein. Starting with the first commercial applications in 2001, D-amps have used in many successful devices.

However, why they are so efficient is not clear. In this chapter, we provide a possible theoretical explanation for this efficiency.

We will do it in two sections. In the first of these sections, we explain why using pulses makes sense. In the following section, we explain why the pulse's duration should linearly depend on the amplitude of the input signal $x(t)$.

8.3   Why Pulses

Let us start with some preliminaries.

There are bounds on possible values of the signal. For each device, there are bounds $\underline{s}$ and $\overline{s}$ on the amplitude of a signal that can be represented by this device. In other words, for each signal $s(t)$ represented by this device and for each moment of time t, we have $\underline{s} \leq s(t) \leq \overline{s}$.

Noise is mostly additive. In every audio system, there is noise $n(t)$. Most noises are *additive*: they add to the signal. For example, if we want to listen to a musical performance, then someone talking or coughing or shuffling in a chair produce a noise signal that adds to the original signal.

Similarly, if we have a weak signal $s(t)$ which is being processed by an electronic system, then additional signals from other electric and electronic devices act as an additive noise $n(t)$, so the actual amplitude is equal to the sum $s(t) + n(t)$.

What do we know about the noise. In some cases—e.g., in a predictable industrial environment—we know what devices produce the noise, so we can predict some statistical characteristics of this noise.

However, in many other situations—e.g., in TV sets—noises are rather unpredictable. All we may know is the upper bound n on possible values of the noise $n(t)$: $|n(t)| \leq n$.

We want to be the least vulnerable to noise. To improve the quality of the resulting signal, it is desirable to select a signal's representation in which this signal would be the least vulnerable to noise.

Additive noise $n(t)$ changes the original value $s(t)$ of the signal to the modified value $s(t) + n(t)$. Ideally, we want to make sure that, based on this modified signal, we will be able to uniquely reconstruct the original signal $s(t)$.

For this to happen, not all signal values are possible. Let us show that for this desired feature to happen, we must select a representation $s(t)$ in which not all signal values are possible.

Indeed, if we have two possible values $s_1 < s_2$ whose difference does not exceed $2n$, i.e., for which $\dfrac{s_2 - s_1}{2} \leq n$, then we have a situation in which:

- to the first value s_1, we add noise $n_1 = \dfrac{s_2 - s_1}{2}$ for which $|n_1| \leq n$, thus getting the value $s_1 + n_1 = \dfrac{s_1 + s_2}{2}$; and

- to the second value s_2, we add noise $n_2 = -\dfrac{s_2 - s_1}{2}$ for which also $|n_2| \leq n$, thus also getting the same value $s_2 + n_2 = \dfrac{s_1 + s_2}{2}$.

Thus, based on the modified value $s_i + n_i = \dfrac{s_1 + s_2}{2}$, we cannot uniquely determine the original value of the signal: it could be s_1 or it could be s_2.

Resulting explanation of using pulses. The larger the noise level n against which we are safe, the larger must be the difference between possible values of the signal. Thus, to makes sure that the signal representation is safe against the strongest possible noise, we must have the differences as large as possible.

On the interval $[\underline{s}, \overline{s}]$, the largest possible difference between the possible values is when one of these values is $\underline{s}$ and another value is $\overline{s}$. Thus, we end up with a representation in which at each moment of time, the signal is equal either to $\underline{s}$ or to $\overline{s}$.

This signal can be equivalently represented as a constant level $\underline{s}$ and a sequence of pulses of the same height (amplitude) $\overline{s} - \underline{s}$.

8.4 Why the Pulse's Duration Should Linearly Depend on the Amplitude of the Input Signal

Pulse representation: reminder. As we concluded in the previous section, to make the representation as noise-resistant as possible, we need represent the time-dependent input signal $x(t)$ by a sequence of pulses.

How can we encode the amplitude of the input signal. Since the height (amplitude) of each pulse is the same, the only way that we can represent information about the input signal's amplitude $x(t)$ at a given moment of time t is to make the pulse's width (duration) w—and the time interval between the pulses—dependent on $x(t)$. Thus, we should select w depending on the amplitude $w = w(x)$.

Which dependence $w = w(x)$ should we select: analysis of the problem. Which dependence of the width w on the initial amplitude should we select?

Since the amplitude x of the input is encoded by the pulse's width $w(x)$, the only way to reconstruct the original signal x is to measure the width w, and then find x for which $w(x) = w$.

We want this reconstruction to be as accurate as possible. Let $\varepsilon > 0$ be the accuracy with which we can measure the pulse's duration w. A change Δx in the input signal—after which the signal becomes equal to $x + \Delta x$—leads to the changed width $w(x + \Delta x)$. For small Δx, we can expand the expression $w(x + \Delta x)$ in Taylor series and keep only linear terms in this expansion:

$$w(x + \Delta x) \approx w(x) + \Delta w,$$

where we denoted $\Delta w \stackrel{\text{def}}{=} w'(x) \cdot \Delta x$. Differences in width which are smaller than ε will not be detected. So the smallest difference in Δx that will be detected comes from the formula $|\Delta w| = |w'(x) \cdot \Delta x| = \varepsilon$, i.e., is equal to $|\Delta x| = \dfrac{\varepsilon}{|w'(x)|}$.

In general, the guaranteed accuracy with which we can determine the signal is thus equal to the largest of these values, i.e., to the value

$$\delta = \max_x \frac{\varepsilon}{|w'(x)|} = \frac{\varepsilon}{\min_x |w'(x)|}.$$

We want this value to be the smallest, so we want the denominator $\min_x |w'(x)|$ to attain the largest possible value.

The limitation is that the overall widths of all the pulses corresponding to times

$$0 = t_1, t_2 = t_1 + \Delta t, \ldots, t_n = T$$

should fit within time T, i.e., we should have

$$\sum_{i=1}^{n} w(x(t_i)) \leq T.$$

This inequality must be satisfied for all possible input signals. In real life, everything is bounded, so at each moment of time, possible values $x(t)$ of the input signal must be between some bounds $\underline{x}$ and $\overline{x}$: $\underline{x} \leq x(t) \leq \overline{x}$.

We want to be able to uniquely reconstruct x from $w(x)$. Thus, the function $w(x)$ should be monotonic—either increasing or decreasing.

If the function $w(x)$ is increasing, then we have $w(x(t)) \leq w(\overline{x})$ for all t. Thus, to make sure that the above inequality is satisfied for all possible input signals, it is sufficient to require that this inequality is satisfied when $x(t) = \overline{x}$ for all t, i.e., that $n \cdot w(\overline{x}) \cdot T$, or, equivalently, that $w(\overline{x}) \leq \Delta t = \dfrac{T}{n}$.

If the function $w(x)$ is decreasing, then similarly, we conclude that the requirement that the above inequality is satisfied for all possible input signals is equivalent to requiring that $w(\underline{x}) \le \Delta t$.

Thus, to find the best dependence $w(x)$, we must solve the following optimization problem:

Which dependence $w = w(x)$ should we select: precise formulation of the problem. To find the encoding $w(x)$ that leads to the most accurate reconstruction of the input signal $x(t)$, we must solve the following two optimization problems.

- Of all increasing functions $w(x) \ge 0$ defined for all $x \in [\underline{x}, \overline{x}]$ and for which $w(\overline{x}) \le \Delta t$, we must select a function for which the minimum $\min_x |w'(x)|$ attains the largest possible value.
- Of all decreasing functions $w(x) \ge 0$ defined for all $x \in [\underline{x}, \overline{x}]$ and for which $w(\underline{x}) \le \Delta t$, we must select a function for which the minimum $\min_x |w'(x)|$ attains the largest possible value.

Solution to the above optimization problem. Let us first show how to solve the problem for the case when the function $w(x)$ is increasing. We will show that in this case, the largest possible value of the minimum $\min_x |w'(x)|$ is attained for the function $w(x) = \Delta t \cdot \dfrac{x - \underline{x}}{\overline{x} - \underline{x}}$.

Indeed, in this case, the dependence $w(x)$ on x is linear, so we have the same value of $w'(x)$ for all x—which is thus equal to the minimum:

$$\min_x |w'(x)| = w'(x) = \frac{\Delta t}{\overline{x} - \underline{x}}.$$

Let us prove that the minimum cannot attain any larger value. Indeed, if we could have

$$m \overset{\text{def}}{=} \min_x |w'(x)| > \frac{\Delta t}{\overline{x} - \underline{x}},$$

then we will have, for each x,

$$w'(x) \ge m > \frac{\Delta t}{\overline{x} - \underline{x}}.$$

Integrating both sides of this inequality by x from $\underline{x}$ to $\overline{x}$, and taking into account that

$$\int_{\underline{x}}^{\overline{x}} w'(x)\, dx = w(\overline{x}) - w(\underline{x})$$

and

$$\int_{\underline{x}}^{\overline{x}} m\, dx = m \cdot (\overline{x} - \underline{x}),$$

we conclude that $w(\overline{x}) - w(\underline{x}) \geq m \cdot (\overline{x} - \underline{x})$. Since $m > \dfrac{\Delta t}{\overline{x} - \underline{x}}$, we have $m \cdot (\overline{x} - \underline{x}) > \Delta t$ hence

$$w(\overline{x}) - w(\underline{x}) > \Delta t.$$

On the other hand, from the fact that $w(\overline{x}) \leq \Delta$ and $w(\underline{x}) \geq 0$, we conclude that $w(\overline{x}) - w(\underline{x}) \leq \Delta t$, which contradicts to the previous inequality. This contradiction shows that the minimum cannot attain any value larger than $\dfrac{\Delta t}{\overline{x} - \underline{x}}$, and thus, that the linear function $w(x) = \Delta t \cdot \dfrac{x - \underline{x}}{\overline{x} - \underline{x}}$ is indeed optimal.

Similarly, in the decreasing case, we can prove that in this case, the linear function $w(x) = \Delta t \cdot \dfrac{\overline{x} - x}{\overline{x} - \underline{x}}$ is optimal.

In both cases, we proved that the width of the pulse should linearly depend on the input signal—i.e., that the encoding use in class-D audio amplifiers is indeed optimal.

References

1. Alvarez, K., Urenda, J.C., Kreinovich, V.: Why Class-D audio amplifiers work well: a theoretical explanation. In: Ceberio, M., Kreinovich, V. (eds.) How Uncertainty-Related Ideas Can Provide Theoretical Explanation for Empirical Dependencies, pp. 15–20. Springer, Cham, Switzerland (2021)
2. Santo, B.: Bringing big sound to small devices: One module revolutionized how we listen. IEEE Spectr. **10**, 17 (2019)

Chapter 9
Application to Mechanical Engineering: Wood Structures

Wood is a very mechanically anisotropic material. At each point on the wooden beam, both average values and fluctuations of the local mechanical properties corresponding to a certain direction depend, e.g., on whether this direction is longitudinal, radial or tangential with respect to the grain orientation of the original tree. This anisotropy can be described in geometric terms, if we select a point x and form *iso-correlation surfaces*—i.e., surfaces formed by points y with the same level of correlation $\rho(x, y)$ between local changes in the vicinities of the points x and y. Empirical analysis shows that for each point x, the corresponding surfaces are well approximated by concentric homothetic ellipsoids. In this chapter, we provide a theoretical explanation for this empirical fact.

Results from this chapter first appeared in [1].

9.1 Problem: Need for a Theoretical Explanation of an Empirical Fact

Wood is a very mechanically anisotropic material. At each point on the wooden beam, both average values and fluctuations of the local mechanical properties corresponding to a certain direction depend, e.g., on whether this direction is longitudinal, radial or tangential with respect to the grain orientation of the original tree. This anisotropy can be described in geometric terms, if we select a point x and form *iso-correlation surfaces*—i.e., surfaces formed by points y with the same level of correlation $\rho(x, y)$ between local changes in the vicinities of the points x and y.

Empirical analysis shows that for each point x, the corresponding surfaces are well approximated by concentric homothetic ellipsoids; see, e.g., [2]. How can we explain this empirical fact?

© The Author(s), under exclusive license to Springer Nature Switzerland AG 2022
J. C. Urenda and V. Kreinovich, *Algebraic Approach to Data Processing*,
Studies in Big Data 115, https://doi.org/10.1007/978-3-031-16780-5_9

41

9.2   Our Explanation: Main Idea

In this chapter, we provide a theoretical explanation for this empirical fact. The main ideas behind our explanation are similar to the ideas used in [3, 4] to explain efficiency of ellipsoid approximation in numerical analysis (see, e.g., [5–14]); the main difference is that now we consider:

- *not* classes of sets (such as the class of all ellipsoids), but
- classes of *families* of sets (e.g., the class of all families of concentric homothetic ellipsoids).

Specifically, we show that for the smallest dimension d for which it is possible to have an affine-invariant optimality criterion on the space of all such d-dimensional classes, for any such criterion, the optimal family consists of concentric homothetic ellipsoids. Thus, such families of ellipsoids provide the optimal approximation to the actual surfaces—at least in the *first* approximation, i.e., approximation corresponding to the smallest possible number of parameters.

9.3   Our Explanation: Details

Family of sets: towards a precise definition. For each spatial point x, we would like to describe, for each possible value ρ_0 of the correlation $\rho(x, y)$, the set $S_{\rho_0}(x)$ of all the points y for which the correlation $\rho(x, y)$ between the values at x and y is greater than or equal to ρ_0.

What are the natural properties of these families of sets?

First property: coverage. For each y, there is some value of $\rho(x, y)$, so for this x, the union of all these sets $S_{\rho_0}(x)$ coincides with the whole space.

Second property: monotonicity. Of course, if $\rho(x, y) \geq \rho_0$ and $\rho_0 \geq \rho_0'$, then $\rho(x, y) \geq \rho_0'$. So, the sets $S_{\rho_0}(x)$ should be inclusion-monotonic: if $\rho_0 \leq \rho_0'$, then $S_{\rho_0'}(x) \subseteq S_{\rho_0}(x)$.

Third property: boundedness. From the physical viewpoint, the further away is the point y from the point x, the less the physical quantities corresponding to these points are correlated. As the distance increases, this correlation should tend to 0. Thus, each set $S_{\rho_0}(x)$ is *bounded*.

Fourth property: continuity. In physics, most processes are continuous—with the exception of processes like fracturing, which we do not consider here. We can therefore conclude that the correlation $\rho(x, y)$ continuously depends on y. So, if we have $\rho(x, y_n) \geq \rho_0$ for some sequence of points y_n that converges to a point y ($y_n \to y$), then we should have $\rho(x, y) = \lim\limits_{n \to \infty} \rho(x, y_n) \geq \rho_0$. In other words, if $y_n \in S_{\rho_0}(x)$ and $y_n \to y$, then $y \in S_{\rho_0}(x)$, i.e., each set $S_{\rho_0}(x)$ is *closed*.

Similarly, it is reasonable to conclude that the set $S_{\rho_0}(x)$ should continually depend on ρ_0, i.e., that of the two values ρ_0 and ρ_0' are close, then the corresponding sets $S_{\rho_0}(x)$ and $S_{\rho_0'}(x)$ should also be close. A natural way to describe closeness between (bounded closed) sets is to use the so-called Hausdorff distance. In precise terms, for any $\varepsilon > 0$, we say that the sets A and B are *ε-close* if:

- every point $a \in A$ is ε-close to some point $b \in B$ (in the sense that the distance $d(a, b)$ does not exceed ε: $d(a, b) \leq \varepsilon$), and
- every point $b \in B$ is ε-close to some point $a \in A$.

The Hausdorff distance $d_H(A, B)$ between the sets A and B is then defined as the smallest ε for which the sets A and B are ε-close. It can be shown that this distance can be equivalently defined as follows:

$$d_H(A, B) = \max\left(\sup_{a \in A} d(a, B), \sup_{b \in B} d(b, A)\right),$$

where $d(a, B) \stackrel{\text{def}}{=} \inf\limits_{b \in B} d(a, b)$.

Fifth property: what is the set of possible values of the parameter? In this family of sets, correlation value is a parameter. What are the possible values of correlation? In general, correlations can take any value from -1 to 1. When $y = x$, the correlation is clearly equal to 1. When $y \to \infty$, we get values close to 0. Since the function $\rho(x, y)$ is continuous, this function takes all intermediate values. So, the possible values of the correlation form some interval. In some cases, we may have all possible negative values, in other cases, only some negative values, in yet other cases, we only have non-negative values. So, in general, we will consider all possible intervals of possible value of ρ_0. This interval may be closed—if there are points with - correlation, or is can be open.

Resulting definition. So, we arrive at the following definition.

Definition 9.1 Let $N \geq 2$ be an integer. By a *family of sets*, we mean a set $\{S_c : c \in I\}$ of bounded closed sets $S_c \subseteq \mathbb{R}^N$ obtained by applying, to each real number c from a non-degenerate interval I (open or closed, finite or infinite), a mapping $c \to S_c$ that has the following properties:

- the dependence of S_c on c is continuous: if $c_n \to c$, then $d_H(S_{c_n}, S_c) \to 0$,
- the family S_c is monotonic: if $c < c'$, then $S_{c'} \subseteq S_c$, and
- the union of all the sets S_c coincides with the whole space.

Comments.

- According to this definition, the family remains the same if we simply re-parameterize the family: e.g., if instead of the original parameter c, we use a new parameter $c' = c + c_0$ or $c' = \lambda \cdot c$ for some constants c_0 and λ.
- In our specific problem, we are interested in the 3-D case $N = 3$. However, since we can envision similar problem in the plane $N = 2$ or in higher-dimensional spaces—and since the proof of our main result does not depend on any specific value N—in this chapter, we consider the general case $N \geq 2$.
- We are specifically interested in *concentric homothetic families of ellipsoids*, i.e., in families of the type $S_c = c \cdot E + a$, where a is a given vector, and E is an ellipsoid with a center at 0.

Class of families of sets. For different situation, in general, we get different correlations and thus, different families of sets. We would like to find general class of such families that would, ideally, cover all such situations. We can use different parameters to differentiate different families from this class. In other words, a class can be described as a method for assigning, to each possible combination of values of these parameters, some family. As before, it makes sense to require that the resulting mapping is continuous.

Definition 9.2 Let $N \geq 2$ and $r > 0$ be integers. By a *r-parametric class of families of sets*, we mean a mapping that assigns, to each element $p = (p_1, \ldots, p_r)$ from an open r-dimensional set $D \subseteq \mathbb{R}^r$, a family $\{S_c(p)\}$ so that the dependence of $S_c(p)$ on c and p is continuous.

For our problem, an optimality criterion must be affine-invariant. In our case, we want to compare different classes (of families of sets). In selecting optimality criteria, it is reasonable to take into account that while we want to deal with sets of points in physical space, from the mathematical viewpoint, we deal with sets of tuples of real numbers. Real numbers (coordinates) describing each point depend on what coordinate system we use: if we select a different starting point, then all the coordinates are shifted $x_i \rightarrow x_i + a_i$; if we select different axes for the coordinates, we get a rotation $x_i \rightarrow \sum_{j=1}^{N} r_{ij} \cdot x_j$ for an appropriate matrix r_{ij}, etc.

These transformations make sense for the *isotropic* case, when all the properties of a material are the same in all directions. Wood is an example of an *anisotropic* material: e.g., it is easier to cut it along the orientation of the original tree than across that orientation. It is known that in many cases, the description of an anisotropic material can be reduced to the isotropic case if we apply an appropriate affine transformation. This usually comes from the fact that, e.g., mechanical properties of a body can be described by a symmetric matrix, and each symmetric matrix can become a unit matrix if we use its eigenvectors as the base for the new coordinate system.

In view of this, it is reasonable to require that our optimality criterion is invariant not only with respect to shifts and rotations, but also with respect to all possible affine (linear) transformations. Thus, we arrive at the following definitions.

Definition 9.3 Let $N > 2$ be an integer. By an *affine transformation*, we mean a transformation $T : \mathbb{R}^N \to \mathbb{R}^N$ of the type $(Tx)_i = a_i + \sum_{j=1}^{N} b_{ij} \cdot x_j$ for some reversible matrix b_{ij}. Let T be an affine transformation.

- Let $S \subseteq \mathbb{R}^N$ be a set. By the *result* $T(S)$ of applying T to S, we mean the set $\{T(s) : s \in S\}$.
- Let $F = \{S_c : c \in I\}$ be a family of sets. By the *result* $T(F)$ of applying T to F, we mean the family $\{T(S_c) : c \in I\}$.
- Let $C = \{S_c(p)\}$ be class of families. By the *result* $T(C)$ of applying T to C, we mean the class $\{T(S_c(p))\}$.

Proposition 9.1 *Let $N > 0$ and $r > 0$ an integers, and let $(<, \sim)$ be a final affine-invariant optimality criterion on the set of all r-parametric classes of families of sets in $\mathbb{R}^N$. Then:*

- $r \geq \dfrac{N \cdot (N + 3)}{2} - 1$*; and*
- *for* $r = \dfrac{N \cdot (N + 3)}{2} - 1$*, the optimal class consists of concentric homothetic families of ellipsoids.*

Comment. This result indeed shows that the class of concentric homothetic families of ellipsoids is the simplest of all possible optimal classes—simplest in the sense that it requires the smallest number of parameters to describe.

9.4 Proof

$1°$. As we have shown in Chap. 3, since the optimality criterion is final and affine-invariant, the optimal class is also affine-invariant: with each family F this class also contains the family $T(F)$. This means, in its turn, that for each set S_c from each family, some family from the optimal class also contains the set $T(S_c)$.

$2°$. Let us show that $r \geq \dfrac{N \cdot (N + 3)}{2} - 1$. Indeed, it is known (see, e.g., [15]) that for every non-degenerate bounded set S (i.e., for every bounded set which is not contained in a proper subspace), among all ellipsoids that contain S, there exists a unique ellipsoid of the smallest volume. It is also known that this correspondence between a set and the corresponding ellipsoid is affine-invariant: if an ellipsoid E corresponds to the set S_c, then, for each affine transformation T, to the set $T(S_c)$ there corresponds the ellipsoid $T(E)$.

It is known that every two ellipsoids can be obtained from each other by an appropriate affine transformation. Thus, the family of all ellipsoids corresponding to all the sets from all the families consists of all the ellipsoids. How many ellipsoids are there? A general ellipsoid can be determine by a quadratic formula $\sum_{ij} a_{ij} \cdot x_i \cdot x_j +$

$\sum_{i=1}^{N} a_i \cdot x_i \leq 1$ for some symmetric matrix a_{ij} and a vector a_i—and it is easy to see
that different combinations of the matrix and the vector lead to different ellipsoids.
We need N values $a_1, \ldots, a_N$ to describe a vector. Out of N^2 elements of the matrix,
we need N values to describe its diagonal values a_{ii} and we need $\dfrac{N^2 - N}{2}$ to describe
non-diagonal elements: we divide by two since the matrix is symmetric $a_{ij} = a_{ji}$.
Thus, overall, we need

$$N + N + \frac{N^2 - N}{2} = \frac{N \cdot (N + 3)}{2}$$

values.

Thus, the set of all ellipsoids is $\dfrac{N \cdot (N + 3)}{2}$-dimensional. Since to each set S_c
from families from the optimal class, we assign an ellipsoid, the dimension of the set
of such sets should also be at least $\dfrac{N \cdot (N + 3)}{2}$-dimensional. These sets are divided
into 1-parametric families, so the dimension r of the class of such families cannot
be smaller than the above dimension minus 1. Thus, indeed,

$$r \geq \frac{N \cdot (N + 3)}{2} - 1.$$

3°. Let us now prove that for the smallest possible dimension $r =$
$r_{\min} \overset{\text{def}}{=} \dfrac{N \cdot (N + 3)}{2} - 1$, all the sets S_c from the each family of the optimal class are
ellipsoids.

In Part 2 of this proof, we showed that each ellipsoid is associated with some set S_c
from one of these families. The unit ball with a center at 0 is clearly an ellipsoid. Let
us consider the set S_c which is associated with this unit ball. A unit ball is invariant
with respect to all the rotations around its center. If the associated set S_c is not equal
to the unit ball, this means that this set is not invariant with respect to at least some
rotations. In other words, the group of all rotations that leave this set invariant is a
proper subgroup of the group of all rotations. This implies that the dimension of this
group is smaller than the dimension of the group of all rotations—and thus, that there
exists at least 1-parametric family $\mathcal{R}$ of rotations R with respect to which the set S_c
is not invariant.

Since the optimal class is affine-invariant, all the sets $R(S_c)$ are also sets from
some family from the optimal class—and for all of them, the same unit ball is the
smallest-volume ellipsoid. Thus, for this particular ellipsoid—the unit ball, we have
at least a 1-dimensional family of sets S_c associated with this same ellipsoid. By
applying a generic affine transformation, we can find a similar at-least-1-dimensional
family of sets corresponding to each ellipsoid. Thus, the dimension of the set of all
sets S_c is at least one larger than the dimension of the family of all ellipsoids, i.e.,
at least $\dfrac{N \cdot (N + 3)}{2} + 1 = r_{\min} + 2$. However, we have a $r_{\min}$-dimensional class of

1-dimensional families of sets, so the overall dimension of the set of all the sets S_c cannot be larger than $r_{min} + 1$. This contradiction shows that the set S_c cannot be different from the enclosing minimal-volume ellipsoid. Thus, indeed, each set from each family from the optimal class is an ellipsoid.

$4°$. To complete the proof, we need to prove that ellipsoids in each family are concentric and homothetic.

We have proven that each ellipsoid appears as an appropriate smallest-volume set. Now that we know that each set S_c coincides with its smallest-volume enclosure, we can thus conclude that each ellipsoid appears as one of the sets S_c from one of the families from the optimal class. Similarly to Part 3 of this proof, let us consider the unit ball centered at 0. If the 1-dimensional family F_0 containing this ball is not invariant with respect to all possible rotations around the ball's center, then we have at least a 1-dimensional group of different families containing the same ellipsoid—the unit ball. However, the only way for an r_{min}-dimensional class of 1-dimensional families to cover the whole $(r_{max} + 1)$-dimensional family of ellipsoids is when all elements of all families are different. So we cannot have several families containing the same ellipsoid.

This argument shows that the family F_0 containing the unit ball *should be* rotation-invariant. Since all the sets from this family are included in each other and thus, cannot be transformed into each other by rotations—this means that each ellipsoid from this family F_0 must be rotation-invariant. This means that each ellipsoid from this family must be a ball concentric with our selected unit ball—and thus, homothetic to this ball.

For any other family F, by selecting any ellipsoid E from this family and applying the affine transformation that transforms the above unit ball into E, we get a new family $T(F_0)$ of concentric homothetic ellipsoids. Since an ellipsoid can only belong to one family, we thus conclude that the family F also consists of concentric homothetic ellipsoids.

The proposition is proven.

References

1. Schietzold, F.N., Urenda, J., Kreinovich, V., Graf, W., Kaliske, M.: Why ellipsoids in mechanical analysis of wood structures. In: Proceedings of the 9th International Workshop on Reliable Engineering Computing REC'2021, Taormina, Italy, pp. 604–614 (2021)
2. Schietzold, F.N., Graf, W., Kaliske, M.: Optimization of tree trunk axes locations in polymorphic uncertain modeled timber structures. In: Beer, M., Zio, E. (eds.) Proceedings of the 29th European Safety and Reliability Conference ESREL'2019, pp. 3201–3208. Hannover, Germany, Research Publishing, Singapore (2019)
3. Finkelstein, A., Kosheleva, O., Kreinovich, V.: Astrogeometry, error estimation, and other applications of set-valued analysis. ACM SIGNUM Newslett. **31**(4), 3–25 (1996)
4. Li, S., Ogura, Y., Kreinovich, V.: Limit Theorems and Applications of Set Valued and Fuzzy Valued Random Variables. Kluwer Academic Publishers, Dordrecht (2002)

5. Belforte, G., Bona, B.: An improved parameter identification algorithm for signal with unknown-but-bounded errors. In: Proceeding of the 7th IFAC Symposium on Identification and Parameter Estimation. York, U.K. (1985)
6. Chernousko, F.L.: Estimation of the Phase Space of Dynamic Systems. Nauka publ, Moscow (1988)
7. Chernousko, F.L.: State Estimation for Dynamic Systems. CRC Press, Boca Raton, Florida (1994)
8. Filippov, A.F.: Ellipsoidal estimates for a solution of a system of differential equations. Int. Comput. 2(4), 6–17 (1992)
9. Fogel, E., Huang, Y.F.: On the value of information in system identification. Bounded noise case. Automatica 18, 229–238 (1982)
10. Norton, J.P.: Identification and application of bounded parameter models. In: Proceeding of the 7th IFAC Symposium on Identification and Parameter Estimation. York, U.K. (1985)
11. Schweppe, F.C.: Recursive state estimation: unknown but bounded errors and system inputs. IEEE Trans. Automat. Control 13, 22 (1968)
12. Schweppe, F.C.: Uncertain Dynamic Systems. Prentice Hall, Englewood Cliffs, New Jersey (1973)
13. Soltanov, S.T.: Asymptotics of the function of the outer estimation ellipsoid for a linear singularly perturbed controlled system. In: S. P. Shary and Yu. I. Shokin (eds.), Interval Analysis, Krasnoyarsk, Academy of Sciences Computing Center, Technical Report No. 17, pp. 35–40 (1990)
14. Utyubaev, G.S.: On the ellipsoid method for a system of linear differential equations. In: S. P. Shary (ed.), Interval Analysis, Krasnoyarsk, Academy of Sciences Computing Center, Technical Report No. 16, pp. 29–32 (1990)
15. Busemann, H.: The Geometry of Geodesics. Academic Press, New York (1955)

Chapter 10
Medical Application: Prevention

In this and following chapters, we show that algebraic approach can help on all stages of medicine: prevention, testing, diagnosis, and treatment.

In this chapter, we deal with prevention. Results from the first two chapters first appeared in [1].

10.1 Problem: How to Best Maintain Social Distance

This problem is related to the pandemic-related need to observe a social distance of at least 2 m (6 feet) from each other.

Two persons are on two sides of a narrow-walkway street, waiting for the green light. They start walking from both sides simultaneously. For simplicity, let us assume that they walk with the same speed.

If they follow the shortest distance path—i.e., a straight line connecting their initial locations A and B—they will meet in the middle, which is not good. So one of them should move somewhat to the left, another somewhat to the right. At all moments of time, they should be at least 2 m away from each other. What is the fastest way for them to do it?

10.2 Towards Formulating This Problem in Precise Terms

The situation is absolutely symmetric with respect to the reflection in the midpoint M of the segment AB. So, it is reasonable to require that the trajectory of the second person can be obtained from the trajectory of the first person by this reflection. Thus, at any given moment of time, the midpoint M is the midpoint between the two

J. C. Urenda and V. Kreinovich, *Algebraic Approach to Data Processing*,
Studies in Big Data 115, https://doi.org/10.1007/978-3-031-16780-5_10

persons. In these terms, the requirement that they are separated by at least 2 m means that each of them should always be at a distance at least 1 m from the midpoint M. In other words, both trajectories should avoid the disk of radius 1 meter with a center at the midpoint M.

We want the fastest possible trajectory. Since the speed is assumed to be constant, this means that they should follow the shortest possible trajectory. In other words, we need to find the shortest possible trajectory going from point A to point B that avoids the disk centered at the midpoint M of the segment AB.

10.3 Solution

To get the shortest path, outside the disk, the trajectory should be straight, and where it touches the circle, it should be smooth. Thus, the solution is as follows:

- first, we follow a straight line until it touches the circle as a tangent,
- then, we follow the circle,
- and finally, we follow the straight line again—which again starts as a tangent to the circle:

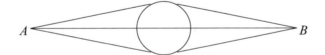

Reference

1. Urenda, J., Kosheleva, O., Ceberio, M., Kreinovich, V.: How mathematics and computing can help fight the pandemic: two pedagogical examples. In: Proceedings of the Annual Conference of the North American Fuzzy Information Processing Society NAFIPS'2020, Redmond, Washington, pp. 20–22 (2020)

Chapter 11
Medical Application: Testing

11.1 Problem: Optimal Group Testing

One of the challenges related to the COVID-19 pandemic is that this disease has an unusually long incubation period—about 2 weeks. As a result, people with no symptoms may be carrying the virus and infecting others. As of now, the only way to prevent such infection is to perform massive testing of the population. The problem is that there is not enough test kits to test everyone.

11.2 What Was Proposed

To solve this problem, researchers proposed the following idea [1, 2]: instead of testing everyone individually, why not combine material from a group of several people and test each combined sample by using a single test kit. If no viruses are detected in the combined sample, this means that all the people from the corresponding group are virus-free, so there is no need to test them again. After this, we need to individually test only folks from the groups that showed the presence of the virus.

11.3 Resulting Problem

Suppose that we need to test a large population of N people. Based on the previous testing, we know the proportion p of those who have the virus. In accordance with the above idea, we divide N people into groups. The question is: what should be the size s of each group?

If the size is too small, we are still using too many test kits. If the size is too big, every group, with a high probability, has a sick person, so we are not dismissing any

© The Author(s), under exclusive license to Springer Nature Switzerland AG 2022
J. C. Urenda and V. Kreinovich, *Algebraic Approach to Data Processing*,
Studies in Big Data 115, https://doi.org/10.1007/978-3-031-16780-5_11

people after such testing, and thus, we are not saving any testing kits at all. So what is the optimal size of the group?

Comment. Of course, this is a simplified formulation, it does not take into account that for large group sizes s, when each individual testing material is diluted too much, tests may not be able to detect infected individuals.

11.4   Let Us Formulate This Problem in Precise Terms

If we divide N people into groups of s persons each, we thus get N/s groups.

The probability that a person is virus-free is $1 - p$. Thus, the probability that all s people from a group are virus-free is $(1 - p)^s$. So, out of N/s groups, the number of virus-free groups is $(1 - p)^s \cdot (N/s)$. Each of these groups has s people, so the overall number of tested people can be obtained by multiplying the number of virus-free groups by s, resulting in $(1 - p)^s \cdot N$. For the remaining $N - (1 - p)^s \cdot N$ folks, we need individual testing. So, the overall number of needed test kits is

$$N_t = \frac{N}{s} + N - (1 - p)^s \cdot N. \tag{11.1}$$

We want to minimize the number of test kits, i.e., we want to find the group size s for which the number (11.1) is the smallest possible.

11.5   Solution

Differentiating the expression (11.1) with respect to s, equating the derivative to 0, and dividing both sides of the resulting equality by N, we get

$$-\frac{1}{s^2} - (1 - p)^s \cdot \ln(1 - p) = 0. \tag{11.2}$$

For small p, we have $(1 - p)^s \approx 1$ and $\ln(1 - p) \approx -p$, so the formula (11.2) takes the form $-\frac{1}{s^2} + p \approx 0$, i.e.,

$$s \approx \frac{1}{\sqrt{p}}. \tag{11.3}$$

For example, for $p = 1\%$, we have $s \approx 10$; for $p = 0.1\%$, we get $s \approx 30$; and for $p = 0.01\%$, we get $s \approx 100$.

The resulting number of tests (11.1) can also be approximately estimated. When the group size s is described by the approximate formula (11.3), we have $\frac{N}{s} \approx$

$\sqrt{p} \cdot N$. If we take into account that $(1 - p)^s \approx 1 - p \cdot s$, then

$$N - (1 - p)^s \cdot N \approx p \cdot s \cdot N \approx \sqrt{p} \cdot N.$$

Thus, we get

$$N_t \approx \sqrt{p} \cdot N. \tag{11.4}$$

For example, for $p = 1\%$, we need 10 times fewer test kits than for individual testing; for $p = 0.1\%$, we need 30 times fewer test kits; and for $p = 0.01\%$, we need 100 times fewer test kits.

References

1. Perry, T.S.: Researchers are using algorithms to tackle the coronavirus test shortage: the scramble to develop new test kits that deliver faster results. IEEE Spect. **57**(6), 4 (2020)
2. Shental, N., Levy, S., Wuvshet, V., Skorniakov, S., Shemer-Avni, Y., Porgador, A., Hertz, T.: Efficient High Throughput SARS-CoV-2 Testing to Detect Asymptomatic Carriers. https://doi.org/10.1101/2020.04.14.20064618, posted on 20 Apr. 2020

Chapter 12
Medical Application: Diagnostics, Part 1

To adequately treat different types of lung dysfunctions in children, it is important to properly diagnose the corresponding dysfunction, and this is not an easy task. Neural networks have been trained to perform this diagnosis, but they are not perfect in diagnostics: their success rate is 60%. In this chapter, we show that by selecting an appropriate invariance-based pre-processing, we can drastically improve the diagnostic success, to 100% for diagnosing the presence of a lung dysfunction.

In this chapter, we explain, in general, how algebraic techniques lead to the corresponding diagnostic procedure. Specific features of this procedure are explained in the following two chapters.

Results from this chapter first appeared in [1].

12.1 Problem: Diagnosing Lung Disfunctions in Children

Lung dysfunctions. One of the major lung dysfunctions is asthma, a long-term inflammatory disease of the airways of the lungs. It is characterized by recurring airflow obstruction, bronchospasms, wheezing, coughing, chest tightness, and shortness of breath. These episodes may occur a few times a day or a few times per week [2].

Asthma may be preceded by Small Airways Impairment (SAI), a chronic obstructive bronchitis. If inflammation persists during SAI, it could cause asthma.

SAI, in its turn, may be preceded by a less severe condition that medical doctors classify as Possible Small Airways Impairment (PSAI).

Diagnostics of different lung dysfunctions is difficult but important. All lung dysfunctions lead to similar symptoms like wheezing, coughing, etc. As a result, it is difficult to distinguish between these dysfunctions—and it is also difficult to distinguish these chronic dysfunctions from a common short-term respiratory disease.

J. C. Urenda and V. Kreinovich, *Algebraic Approach to Data Processing*,
Studies in Big Data 115, https://doi.org/10.1007/978-3-031-16780-5_12

However, the diagnosing of these diseases is very important, because in general, for different diseases, different treatments are efficient.

How different dysfunctions are diagnosed now. Since it is difficult to diagnose different dysfunctions solely based on the symptoms, the corresponding diagnostics involves measuring airflow in different situations. The most effective diagnostic comes from active measurements—spirometry. A patient is asked to deeply inhale, to hold their breath, and to exhale as fully as possible—and the corresponding instrument is measuring the airflow following all these instructions. Based on these measurements, symptoms, and clinical history, medical doctors come up with a diagnosis of different dysfunctions.

Children diagnostics: a serious problem. Unfortunately, the spirometry technique described above does not work with little children, especially children of pre-school age, since it not easy to make them follow the corresponding instructions; see, e.g., [3]. The same problem occurs with elderly patients and patients with certain limitations.

An additional problem is that even when children follow instructions during the spirometry testing, spirometry results are not sensitive enough to detect obstruction of small airways (2 mm or less in diameter); see, e.g., [4–8].

How can we diagnose children: main idea of the corresponding measurements. Since we cannot use active measuring techniques, techniques that require children's active participation, we have to rely on passive techniques, i.e., techniques that do not require such participation. What we can do is change the incoming airflow and measure how that affects the outcoming airflow.

Passive measurements: details. The easiest way of changing the airflow is to switch a certain extra amount of airflow on or off. This is the main idea behind the Impulse Oscillometry System (IOS); see, e.g., [4]. In a usual IOS, the additional airflow is switched on and off with a period of 5 Hz, meaning that we have a 0.1 s period with extra flow, 0.1 s period without, then again a 0.1 s period with extra flow, etc. The system then measures the resulting outflow $y(t)$.

In real-life clinical environment, the measurement result are affected by noise. Because of this noise, the measured values $\widetilde{y}(t)$ somewhat deviate from the actual (unknown) flow results $y(t)$. The deviations $\widetilde{y}(t) - y(t)$ measured at different moments of time t are usually caused by different factors and are, thus, statistically independent. As a result of this independence, the noise is heavily oscillating—i.e., changes with high frequency. To decrease the effect of this noise, it is therefore reasonable to take a Fourier transform and ignore high-frequency components of this transform—since these components are heavily corrupted by noise.

Since the input signal is periodic, with the period of 5 Hz, we expect the output signal to also be periodic, with the same frequency. In general, when we perform Fourier transform on a signal which is periodic with frequency f, we only get components corresponding to multiples of f, i.e., to f, $2f$, $3f$, etc. In our case, this means that we will have components corresponding to 5, 10 Hz, etc. In practice, it was discovered that components above 25 Hz are too noisy to be useful—actually, the most informative values are one corresponding to 5-15 Hz range. The values corresponding to 20 and 25 Hz are also useful, but they are somewhat less informative that the 5–15 Hz values. So, the system returns the components corresponding to 5, 10, 15, 20, and 25 Hz. To be on the safe side, the system also returns the component corresponding to 35 Hz, which sometimes adds some additional information.

Another problem is related to the fact that while it is relatively easy to implement the on-off switching of the input airflow, the actual values of the on- and off-case airflows may change with time: the pressure in the system may decrease, etc. In principle, it is possible to maintain the exact airflow values, but this will make the system too complicated and thus, too expensive. Instead, the existing systems rely on the fact that while it is not easy to maintain the input airflow $a(t)$ at some pre-defined level a_0, it is possible to accurately measure this airflow.

The added airflow $a(t) - a_0$ is relatively small, so when estimating the reaction $y(t)$ of a human breathing system to this airflow, we can safely ignore terms which are quadratic and of higher order in terms of $a(t) - a_0$ and conclude that the dependence is linear: $y(t) = \int c(t, s) \cdot a(s) \, ds$ for some coefficients $c(t, s)$.

The system does not change much during the time when measurements are performed. So if we start the experiment t_0 seconds earlier, i.e., if we take $\widetilde{a}(t) = a(t + t_0)$ instead of $a(t)$, then the output should change accordingly, to $\widetilde{y}(t) = y(t + t_0)$. So, the corresponding linear transformation is shift-invariant. As we have shown in Chap. 3, this means that $y(t)$ is equal to the convolution $y(t) = \int z(t - s) \cdot a(s) \, ds$ of the input $a(t)$ with some function $z(t)$.

It is known that the Fourier transform of the convolution is equal to the product of Fourier transforms. Thus, in this case, for the corresponding Fourier transforms $Z(f)$, $Y(f)$, and $A(f)$, we get $Y(f) = Z(f) \cdot A(f)$. We are interested in the values $Z(f)$ that do not depend on the inputs. In our case, we have computed the values $Y(f)$, so we can also compute the Fourier coefficients $A(f)$, and return the ratios $Z(f) = Y(f)/A(f)$. This is exactly what the IOS system returns: the complex numbers $Z(f) = R(f) + i \cdot X(f)$ that correspond to six frequencies $f = 5, 10, \ldots, 35$. In analogy with electric circuits, the complex value $Z(f)$ is called the *impedance*, its real part $R(f)$ is called *resistance*, and its imaginary part $X(f)$ is called *reactance*.

These six complex numbers—or, equivalently, the six real parts and the six imaginary parts—are what we can use to properly diagnose the lung dysfunction.

It is not easy to make a diagnosis based on IOS data. If we plot the IOS data corresponding to patients with different diagnoses, we see that the corresponding ranges of values $R(f)$ and $X(f)$ have a huge intersection; see, e.g., Figs. 12.1 and 12.2. This shows that it is not easy to diagnose a patient based on IOS data.

How IOS-based diagnosis is performed now. There are no exact formulas that describe the diagnosis based on the six complex values $Z(f)$; the research about the clinical applications of the IOS parameters is still ongoing. However, we do have several patient records for which, on the one hand, we know the corresponding values

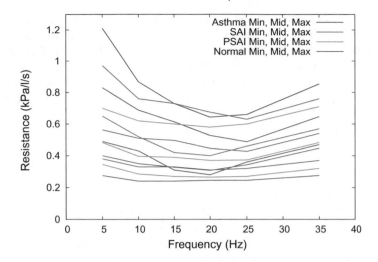

Fig. 12.1 Resistance maximum, middle and minimum patients' curves per class

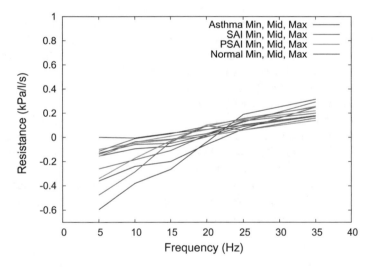

Fig. 12.2 Reactance maximum, middle and minimum patients' curves per class

$Z(f)$, and, on the other hand, we have a diagnosis provided by a skilled medical doctor. It is therefore reasonable to use machine learning and train the system to be able to diagnose a patient.

This has indeed been done: researchers have used either all or some of the 12 numbers (real parts $R(f)$ and imaginary parts $X(f)$) as input and tried to train the neural network to learn the diagnosis in the children patients.

The resulting diagnostic system is, however, not yet perfect. For adult patients, if we use spirometry results as well as IOS, we get an almost perfect separation of asthma from healthy: its accuracy is 98–99% [9, 10]. However, when we only use IOS data, the current system's testing-data accuracy in distinguishing lung dysfunctions such as asthma, SAI, and PSAI from patients who do not have any of these diseases is only close to 60% [11].

Comment. Similar imperfect results were obtained for a related problem: predicting asthma deterioration one week ahead. For this problem, neural networks and other machine learning techniques result, at best, in 70–75% prediction accuracy; see, e.g., [12].

Resulting problem and what we do in this chapter. It is therefore desirable to come up with better diagnostic techniques. Our approach is to help a neural network by providing an appropriate pre-processing of the inputs data. It turns out that an appropriate pre-processing indeed drastically improves the diagnostic results: for diagnosing the presence of a lung dysfunction, we have 100% accuracy.

12.2   First Pre-processing Stage: Scale-Invariant Smoothing

Need for further de-noising. The above-described filtering out of noisy high-frequency components eliminates some noise, but some noise remains.

Smoothing as a way to de-noise. To further decrease the noise level, it is desirable to take into account that in real life, almost all dependencies (including the dependence of the signal intensity on frequency) are smooth—in the sense that a small change in frequency leads to a small change in intensity.

How to smooth a signal. Thus, instead of considering the original (noisy) six complex numbers $Z(f)$ corresponding to $f = 5, 10, \ldots$, it makes sense to approximate these values by a smooth dependence, i.e., by a function of the type $\sum_{j=1}^{k} c_j \cdot e_j(f)$, for some smooth functions $e_1(f), \ldots, e_k(f)$.

Which level of smoothness to choose. Usually, real-life processes are very smooth—with few exceptions like phase transitions. It is therefore desirable to select approximating functions $e_j(f)$ which are as smooth as possible.

   In general, for functions, there are several different degrees of smoothness. The simplest case is when a function is one time differentiable. The next—more smooth—case is when a function is two times differentiable, etc. Then, we have functions which are infinitely many time differentiable, and finally, the smoothest of all—analytical functions, functions that can be expanded in Taylor series. Thus, to achieve maximal smoothness, we will use analytical functions $e_j(f)$.

Which analytical functions should we choose: the idea of scale-invariance. There are many different analytical functions, which ones should we choose?

A natural requirement for this choice comes from the fact that we are approximating a function from numbers (f) to numbers (Z). These numbers represent the values of the corresponding physical quantities. However, the numerical value of each physical quantity depends not only on the quantity itself, it also depends on the measuring unit that we have selected for this quantity. If we replace the original measuring unit with a new one which is λ times smaller, all the numerical values will multiply by λ. In other words, for each frequency, instead of the original numerical value f, we will have a new numerical value $\tilde{f} = \lambda \cdot f$ that describe the exact same physical quantity.

There is no physical reason why some measuring units would be preferable to others. Therefore, it makes sense to require that the selection of the resulting class C of linear combinations $\sum_{j=1}^{k} c_j \cdot e_j(f)$ should not change if we simply re-scale all the values by changing the measuring unit for frequencies. As we have shown in Chap. 3, this means that the functions $e_i(f)$ are linear combinations of the expressions

$$(\ln(f))^p \cdot f^a \cdot (\cos(b \cdot \ln(f)) + \mathrm{i} \cdot \sin(b \cdot \ln(f))).$$

Let us take into account that the functions $e_i(f)$ should be analytical. Now, we can take into account that the functions $e_i(f)$ should be analytical, i.e., they should be expandable in Taylor series for $f = 0$. This requirement excludes possible logarithmic terms $(\ln(f))^p$, as well as cosines and sines of these logarithms, which leaves us with linear combinations of the powers f^a. Due to analyticity, all the powers should be natural numbers, so we conclude that all the functions $e_i(f)$ are linear combinations of expressions $f^0 = 1$, $f^1 = f$, $f^2, \ldots$ In other words, due to scale-invariance, all the functions $e_i(f)$ should be polynomials.

We want to approximate the function $Z(f)$ by a linear combination of the functions $e_i(f)$. A linear combination of polynomials is also a polynomial. Thus, we arrive at the following conclusion.

General conclusion of this section. Due to the natural requirement of scale-invariance, we should approximate the impedance function $Z(f)$ by a polynomial.

12.3 Which Order Polynomials Should We Use?

Formulation of the problem. We want to find the polynomial that fits the observations. Of course, if we take a polynomial of a sufficiently large degree, we can always find a polynomial that fits all observed data exactly—this is a well-known Lagrange interpolation polynomial.

However, the whole purpose of the polynomial smoothing is to de-noise the signal, and if we keep all the values intact, we will retain all the noise. Thus, we should not use polynomials of too high order.

On the other hand, if we use polynomials of too low order—e.g., constants or linear functions—we get a smoothing, but the approximation is too crude, and we lose the information contained in the original signal. How can we find the adequate degree of approximating polynomials?

Natural idea: general case. A natural idea is to take into account the general monotonicity of IOS curves—as described, e.g., in [5]—and select the higher order of approximating polynomials that preserve this monotonicity.

Let us apply this general idea to our problem. According to [5], for all three diseases (asthma, SAI, and PSAI):

- for the real part $R(f)$ of the impedance $Z(f)$, the corresponding value first decreases with frequency f, and then increases;
- for the imaginary part $X(f)$ of the impedance $Z(f)$, the corresponding value increases with frequency f.

So:

- for each degree, we use the usual Least Squares techniques (see, e.g., [13]) to find the polynomial of this degree that best approximates the observed values, and then
- we select the largest degree for which, on the corresponding interval of values of frequency, the resulting best-approximation polynomials follow the same monotonicity pattern.

It turns out that:

- for quadratic and cubic polynomials, we have this feature, but
- for 4th order polynomials, we no longer have the desired monotonicity: for example, for the resistance $R(f)$, the corresponding 4th order polynomial first decreases, then increases, but then decreases again; see, e.g., Fig. 12.3.

Because of this, in this chapter, we approximate the functions $R(f)$ and $X(f)$ by *cubic polynomials*. Let us denote the corresponding approximating polynomials by $R^a(f)$ and $X^a(f)$.

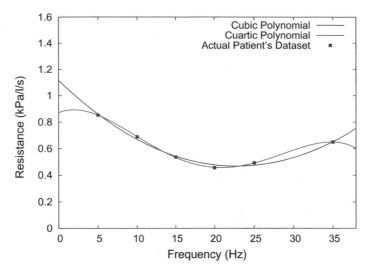

Fig. 12.3 Cubic versus cuartic versus a patient's dataset

12.4   Second Pre-processing Stage: Using the Approximating Polynomials to Distinguish Between Different Diseases

Formulation of the problem. For each diagnosis d, we have several observations corresponding to patients with these diagnosis. Based on these observations, for each patient i with this diagnosis ($i \in d$), we find the smoothed functions $R_i^a(f)$ and $X_i^a(f)$.

Now, when we have a new patient with the corresponding functions $R^a(f)$ and $X^a(f)$, we would like to diagnose this patient, i.e., to classify this patient to one of the groups d. How can we do it?

How to separate different groups: general idea. How do we distinguish groups in general? How do we distinguish cats from dogs? Usually, in such situations:

- we have a mental picture of a typical cat,
- we have a mental picture of a typical dog, and
- we make our decision based on how similar the observed object is to one of these two typical ones.

A natural idea is thus:

- to form, for each group corresponding to a given diagnosis d, "typical" function $R_d(f)$ and $X_d(f)$ corresponding to this diagnosis, and then

- to base our diagnosis of new, yet-undiagnosed patients based on how similar their functions $R^a(f)$ and $X^a(f)$ are to the typical functions corresponding to each diagnosis.

A natural way to form a typical function. A natural way to form the typical function $R_d(f)$ corresponding to a given diagnosis d is to take all the IOS values $R_i(f)$ corresponding to patients i with this diagnosis ($i \in d$), and use Least Squares to find the cubic function $R_d(f)$ that best approximate all these measurement results, i.e., for which the sum $\sum_{i \in d} \sum_f (R_d(f) - R_i(f))^2$ attains its smallest possible value (here, the summation is over the IOS frequencies $f = $ 5, 10, 15, 20, 25, and 35 Hz).

Similarly, a natural way to form the typical function $X_d(f)$ corresponding to a given diagnosis d is to take all the IOS values $X_i(f)$ corresponding to patients i with this diagnosis, and use Least Squares to find the cubic function $X_d(f)$ that best approximate all these measurement results, i.e., for which, the sum $\sum_{i \in d} \sum_f (X_d(f) - X_i(f))^2$ attains its smallest possible value.

12.5 Third Pre-processing Stage: Scale-Invariant Similarity/Dissimilarity Measures

How to describe the similarity of functions: preliminary analysis. To use the above general idea for diagnosing a patient, we need to select a numerical measure of similarity/dissimilarity between the function $R^a(f)$ describing the new patient and the function $R_d(f)$ describing a typical patient with diagnosis d.

One thing to take into account is that usually the measurement record produced by the IOS device contains an initial spike, when the measured value jumps from the original 0 value to a non-zero value corresponding to the actual measurement. This initial impulse-type spike affect the Fourier transform values $R(f)$ and $X(f)$ produced by the measuring device. Since the Fourier transform of an impulse is a constant function, this means that to all the measured values $R_i(f)$ and $X_i(f)$, a constant is added that corresponds to the Fourier transform of this original impulse. Because of this added constant, the same constant gets added to the approximating cubic curves $R_i^a(f)$ and $X_i^a(f)$. For the same patient with the same disease, in different measurements, the initial impulse may be slightly different. Thus, the corresponding added constants may be different for the two measurements of the same patient. In other words, for the same patient, in two consequent measurements, we may get two functions differing by a constant.

So, to properly match, e.g., the patient's function $R^a(f)$ with the function $R_d(f)$ describing a typical patient with diagnosis d, we need to take this possible constant difference into account.

How do we estimate the corresponding constant difference? As we have mentioned earlier, the most informative part of the IOS results correspond to the 5–15 Hz range. The central point of this range is the value 10 Hz. It is therefore reasonable to take the difference $\Delta R_{id} \stackrel{\text{def}}{=} R_d(10) - R_i^a(10)$ of the values corresponding to this central frequency as an estimate for the constant difference. Thus, we should compare the typical function $R_d(f)$ not with the actual patient's function $R_i^a(f)$, but with the "shifted" function $R_i^a(f) + \Delta R_{id}$, shifted so as to provide the best match between the two functions.

In other words, to diagnose a patient i, we need to describe the similarity/dissimilarity between the functions $R_i^a(f) + \Delta R_{id}$ and $R_d(f)$.

Similarly, for reactance, we need to describe the similarity/dissimilarity between the functions $X_i^a(f) + \Delta X_{id}$ and $X_d(f)$, where $\Delta X_{id} \stackrel{\text{def}}{=} X_d(10) - X_i^a(10)$.

How to describe the similarity/dissimilarity of functions: general idea. In general, if we have two functions $F(f)$ and $G(f)$, how can we describe their similarity/dissimilarity? For each frequency f, the larger the absolute value $|F(f) - G(f)|$ of the difference, the less similar are the corresponding values. Thus, it makes sense to assume that the degree of dissimilarity between these two values is a monotonic function of this absolute value: $m(|F(f) - G(f)|)$, for some increasing function $m(x)$.

The overall degree of dissimilarity we can then estimate by simply adding the degree corresponding to different frequencies f, i.e., by considering an integral

$$\int m(|F(f) - G(f)|)\, df.$$

Remaining question. Which similarity/dissimilarity measures—i.e., which functions $m(x)$—should we use?

Main idea: use scale-invariance. To select an appropriate similarity/dissimilarity measure, let us use the same scale-invariance idea that we used to select an approximating family of functions.

Scale-invariance: from idea to formulas. If, for measuring real and imaginary components of the impedance, we select a new unit which is λ times smaller than the original one, then all the corresponding numerical values $F(f)$ and $G(f)$ get multiplied by λ:

- instead of $F(f)$, we get $\widetilde{F}(f) = \lambda \cdot F(f)$, and
- instead of $G(f)$, we get $\widetilde{G}(f) = \lambda \cdot G(f)$.

In this case, the absolutely value of the difference $x = |F(f) - G(f)|$ also gets multiplied by λ:

$$\widetilde{x} = |\widetilde{F}(f) - \widetilde{G}(f)| = |\lambda \cdot F(f) - \lambda \cdot G(f)| = \lambda \cdot |F(f) - G(f)| = \lambda \cdot x.$$

We want to select the function $m(x)$ (that describes degree of similarity/dissimilarity) in such a way that if we change a measuring unit for $F(f)$ and $G(f)$, the resulting value of closeness will not change—provided, of course, we appropriately change a unit for measuring dissimilarity.

As we have shown in Chap. 3, this requirement implies that $m(x) = C \cdot x^{\alpha}$.

By selecting an appropriate unit for measuring dissimilarity, we can make the coefficient C equal to 1. Thus, we arrive at the following conclusion.

Conclusion of this section. Due to scale-invariance, we measure dissimilarity between two functions as

$$\int |F(f) - G(f)|^{\alpha} \cdot df.$$

Remaining question. What value α should we use?

12.6 How to Select α: Need to Have Efficient and Robust Estimates

General idea. In the computer, we can only represent finitely many different values $f_1, f_2, \ldots$ So, an integral, in effect, means a (weighted) sum $\sum |F_i - G_i|^{\alpha}$, where we denoted $F_i \overset{\text{def}}{=} F(f_i)$ and $G_i \overset{\text{def}}{=} G(f_i)$.

From this viewpoint, which value α should we choose?

Need for efficient estimates. We want to be able to have an efficient algorithm that finds the closest approximation, i.e., an approximation for which the dissimilarity degree $\sum |F_i - G_i|^{\alpha}$ is the smallest possible.

It is known (see, e.g., [14, 15]) that, in general, feasible algorithms exist for minimizing convex objective functions, while in many non-convex cases, optimization is NP-hard (i.e., crudely speaking, not feasible). Moreover, it has been proven [16] that, in general, minimization is feasible *only* for convex objective functions. Thus, it makes sense to select the value α in such a way that the objective function $\sum |F_i - G_i|^{\alpha}$ be convex.

In general, according to calculus, a function is convex if its second derivative is non-negative. For $x > 0$, the first derivative of the function x^{α} is $\alpha \cdot x^{\alpha-1}$, and the second derivative is equal to $\alpha \cdot (\alpha - 1) \cdot x^{\alpha-2}$. Here, $\alpha > 0$—the larger the difference, the less similar are the functions, and the value $x^{\alpha-2}$ is also always positive. Thus, the second derivative is non-negative if and only $\alpha - 1 \geq 0$, i.e., if and only if $\alpha \geq 1$.

Thus, to make sure that the corresponding optimization problems can be efficiently solved, we need to select $\alpha \geq 1$.

Need for robustness. Another important requirement for selecting α is to make sure that the resulting estimates are the least affected by noise, i.e., are the most *robust*.

It is known (see, e.g., [17]), that among all the methods based on the objective function $\sum |F_i - G_i|^{\alpha}$ with $\alpha \geq 1$, the most robust is the method corresponding to $\alpha = 1$. Thus, to guarantee the desired robustness, we will use $\alpha = 1$. So, we arrive at the following conclusion.

Conclusion of this section. Among all computationally efficient scale-invariant dissimilarity measures, the most robust (i.e., the most resistant to noise) is the dissimilarity measure

$$\int |F(f) - G(f)| \cdot df.$$

In our case, $F(f) = R_d(f)$ and $G(f) = R_i^a(f) + \Delta R_{id}$, so we need to use the dissimilarity measure

$$\int |R_d(f) - (R_i^a(f) + \Delta R_{id})| \, df.$$

When this dissimilarity measure is close to 0, this means that the function $R_i^a(f)$ corresponding to the i-th patient is very similar to the typical function $R_d(f)$ corresponding to diagnosis d. The more dissimilar these two functions, the larger the value of this dissimilarity measure.

Similarly for reactance, we use the dissimilarity measure

$$\int |X_d(f) - (X_i^a(f) + \Delta X_{id})| \, df.$$

12.7   Scale-Invariance Helps to Take Into Account That Signal Informativeness Decreases with Time

For IOS, the starting part of the signal is more informative. In the previous sections, we implicitly assumed that the values of the signal $y(t)$ at different moments of time are equally informative. However, a typical IOS measurement lasts for 30–45 s—a reasonable time to be tied in to a strange apparatus. As a result, children's level of stress somewhat increases as the measurement process continues. This stress level affects the breathing process—and thus, the measurement results.

So, we can conclude that values $y(t)$ corresponding to earlier time are more informative than values corresponding to later moments of time t.

How can we take this phenomenon into account. A reasonable idea of taking the above phenomenon into account is to consider not the original signals $y(t)$, but the signals weighted with some weight $w(t)$ which decreases with time. In other words, instead of the original signals $y(t)$, we consider weighted signals $w(t) \cdot y(t)$.

Which weight function should we choose: let us again apply scale-invariance. Which weight function $w(t)$ should we choose? A reasonable idea is to again use scale-invariance. In other words, we assume that if we change the unit of time to a one which is λ times smaller—which means changing all numerical values of time from t to $\tilde{t} = \lambda \cdot t$, then the formula for the weight remains the same—once we appropriately re-scale the weight function w as well, from w to $\tilde{w} = \mu(\lambda) \cdot w$, for some function $\mu(\lambda)$.

Again, as we have shown in Chap. 3, this requirement implies that $w = A \cdot t^\alpha$. Since we assume that the weight decreases with time, we must have $\alpha < 0$.

Which value α should we choose? Whether we use the original signal or its weighted form, what we will do next is apply Fourier transform. The original IOS device already returns the Fourier coefficients of the original signal $y(t)$. Thus, from the computational viewpoint, it is desirable to select α for which the Fourier transform of the weighted function $y(t) \cdot t^\alpha$ can be described in terms of the Fourier transform of the original function $y(t)$.

It is known that such a description is possible only for integer values α; namely:

- the Fourier transform of $y(t)/t$ is proportional to the integral of the Fourier transform of $y(t)$;
- the Fourier transform of $y(t)/t^2$ is proportional to the second integral (integral of an integral) of the Fourier transform of $y(t)$; etc.

The simplest of these cases is the case $\alpha = -1$, which corresponds to the integral.

Thus, in addition to the original Fourier transform values, we should consider integrals of these values.

Details. Of course, when computing these integrals, we should take into account the smoothing that we have applied to the original signal. In other words, we should integrate not the original values $Z(f)$, but the corresponding smoothing polynomial approximations.

Integration should be considered over the most informative part of the spectrum— from 5 to 15 Hz. Thus, we arrive at the following conclusion.

Conclusion of this section. In addition to the smoothed signals $R^a(f)$ and $X^a(f)$, we should also consider their integrals $I_R(f) = \int_5^f R^a(x)\,dx$ and $X_R(f) = \int_5^f X^a(x)\,dx$.

12.8 Pre-processing Summarized: What Information Serves as An Input to a Neural Network

Let us summarize. Let us summarize the scale-invariance-motivated pre-processing steps, and thus, describe what inputs are fed into a neural network.

This whole process consists of two stages:

- In the first, preliminary stage, we process data about known patients to find the "typical" functions $R_d(f)$ and $X_d(f)$ that correspond to each diagnosis d.
- On the working stage, we use these typical functions to diagnose a new patient.

Preliminary stage. First, for each diagnosis d, we process patients with known diagnoses d to find the typical functions $R_d(f)$ and $X_d(f)$ corresponding to each of these diagnoses. This is done as follows.

For each patient with the known diagnosis, we get the IOS values $R(f)$ and $X(f)$ corresponding to frequencies f equal to 5, 10, 15, 20, 25, and 35 Hz.

Then, we use the Least Squares techniques to find the coefficients r_0, r_1, r_2, and r_3 of the cubic polynomial

$$R_d(f) = r_0 + r_1 \cdot f + r_2 \cdot f^2 + r_3 \cdot f^3$$

that best approximates the measured values $R_i(f)$ corresponding to the patients with this diagnosis d (of course, we only use patients from the training set, to be able to test our results of the patients from the testing set). In other words, we find the coefficients of the cubic polynomial for which the sum

$$\sum_{i \in d} \sum_f (R_d(f) - R_i(f))^2$$

is the smallest possible.

Similarly, we use the Least Squares techniques to find the coefficients x_0, x_1, x_2, and x_3 of the cubic polynomial

$$X_d(f) = x_0 + x_1 \cdot f + x_2 \cdot f^2 + x_3 \cdot f^3$$

that best approximates the measured values $X_i(f)$ corresponding to all the patients with this diagnosis d. In other words, we find the coefficients of the cubic polynomial for which the sum $\sum_{i \in d} \sum_f (X_d(f) - X_i(f))^2$ is the smallest possible.

For each of these functions, we then compute the integrals $I_{R,d}(f) = \int_5^f R_d(x)\,dx$ and $I_{X,d}(f) = \int_5^f X_d(x)\,dx$.

Working stage. For a new patient, we get the IOS values $R(f)$ and $X(f)$ corresponding to frequencies f equal to 5, 10, 15, 20, 25, and 35 Hz. Then:

- We use the Least Squares techniques to find the coefficients r_0, r_1, r_2, and r_3 of the cubic polynomial

$$R^a(f) = r_0 + r_1 \cdot f + r_2 \cdot f^2 + r_3 \cdot f^3$$

that best approximates the measured values $R(f)$. We also compute the integral $I_R(f) = \int_5^f R^a(x)\,dx$.
- After that, we use the Least Squares techniques to find the coefficients x_0, x_1, x_2, and x_3 of the cubic polynomial

$$X^a(f) = x_0 + x_1 \cdot f + x_2 \cdot f^2 + x_3 \cdot f^3$$

that best approximates the measured values $X(f)$. We also compute the integral $I_X(f) = \int_5^f X^a(x)\,dx$.

- Finally, for each of the four diagnoses d, we compute the following values:

  - $\int_5^{15} |R_d(f) - (R^a(f) + \Delta R_d)|\,df$, where $\Delta R_d \stackrel{\text{def}}{=} R_d(10) - R^a(10)$;
  - $\int_5^{15} |I_{R,d}(f) - (I_R^a(f) + \Delta I_{R,d})|\,df$, where $\Delta I_{R,d} \stackrel{\text{def}}{=} I_{R,d}(10) - I_R^a(10)$;
  - $\int_5^{15} |X_d(f) - (X^a(f) + \Delta X_d)|\,df$, where $\Delta X_d \stackrel{\text{def}}{=} X_d(10) - X^a(10)$;
  - $\int_5^{15} |I_{X,d}(f) - (I_X^a(f) + \Delta I_{X,d})|\,df$, where $\Delta I_{X,d} \stackrel{\text{def}}{=} I_{X,d}(10) - I_X^a(10)$.

These four tuples of four values corresponding to four diagnoses—the total of 16 values—can then be used to train a neural networks to diagnose the patient.

Do we need all these 16 inputs? In data processing, it is known that if we use too many inputs, the prediction accuracy decreases. Indeed, if we use too many inputs, then, together with the most informative ones, we also add less informative ones. These additional inputs add noise to the result of data processing without providing us with any useful information.

Because of this, in data processing in general, it is a good idea not just to use all possible inputs, but also to check if selecting only some of these inputs will leads to more accurate results.

In our case, we tested whether we need both values corresponding to resistance R and values corresponding to reactance X. Interestingly, it turned out that the reactance-related values only decrease the prediction quality. As a result, our recommendation is to only use resistance-related values when diagnosing patients.

12.9   The Results of Training Neural Networks on These Pre-processed Data

Available data. In our research, we used the data collected by our colleague Erika Meraz [18]. This data consists of 288 IOS data sets from patients with known diagnoses.

Pre-processing: first stage. First, for each data set, we used Least Square to find the coefficients of cubic polynomials $R_i^a(f)$ that best fit the observed IOS values $R_i(f)$. Then, we computed the integral $I_{R,i}(f) = \int_5^f R_i^a(x)\,dx$.

Neural network: general idea. We trained a neural network to distinguish patients with lung dysfunctions from patients without lung dysfunction.

A neural networks consist of neurons. Each neuron takes several inputs $x_1, \ldots, x_n$ and transforms them into the signal

$$y = s_0(w_1 \cdot x_1 + \ldots + w_n \cdot x_n - w_0),$$

where $w_0, w_1, \ldots, w_n$ are coefficients that need to be determined during training, and $s_0(z)$ is a nonlinear function known as the *activation function*.

In this research, we used neural networks with sigmoid activation function $s_0(z) = 1/(1 + \exp(-z))$, the most widely used activation function. It is worth mentioning that this function can also be justified by invariance: this time, by shift-invariance (and not scale-invariance, as in the previous examples); see, e.g., [19].

Separation into training and validation data sets. Overall, we had 288 data sets, of which:

- 257 data sets correspond to patients wit lung dysfunctions, and
- 31 data sets correspond to patients without lung dysfunctions.

To train a neural networks, we separated the corresponding data set into training data set (used for training) and validation data set (used for validation). In all three cases, we used approximately 75% of the data for training and approximately 25% for validation; see, e.g., [13, 20]. Specifically:

- we selected 214 data sets for training, among which 191 corresponded to patients with lung dysfunctions and 23 corresponded to patients without lung dysfunctions, and
- the remaining 74 data sets were used for validation, among which 66 corresponded to patients with lung dysfunctions, and 8 patients without lung dysfunctions.

We have four possible diagnoses: asthma (a), SAI (s), PSAI (p), and the absence of lung dysfunctions (n). Within the training set, for each of these four diagnoses d, we applied the Least Square method to all the values $R_i(f)$ corresponding to the data sets with this diagnosis to compute the typical values $R_d(f)$ corresponding to this diagnosis. We then computed the integral $I_{R,d}(f) = \int_5^f R_d(x)\,dx$.

Resulting typical functions $R_d(f)$. As a result of this analysis, we got the following typical functions corresponding to different diagnoses; see Fig. 12.4:

$$R_a(f) = 1.152 - 7.842 \cdot 10^{-2} \cdot f + 2.686 \cdot 10^{-3} \cdot f^2 - 2.443 \cdot 10^{-5} \cdot f^3;$$

$$R_s(f) = 8.960 \cdot 10^{-1} - 5.738 \cdot 10^{-2} \cdot f + 2.067 \cdot 10^{-3} \cdot f^2 - 2.024 \cdot 10^{-5} \cdot f^3;$$

$$R_p(f) = 6.076 \cdot 10^{-1} - 2.717 \cdot 10^{-2} \cdot f + 8.278 \cdot 10^{-4} \cdot f^2 - 4.888 \cdot 10^{-6} \cdot f^3;$$

$$R_n(f) = 4.612 \cdot 10^{-1} - 1.508 \cdot 10^{-2} \cdot f + 4.789 \cdot 10^{-4} \cdot f^2 - 2.424 \cdot 10^{-6} \cdot f^3.$$

Pre-processing: second stage. For each patient i, and for each of the four diagnoses d, we computed the following two similarity/dissimilarity measures:

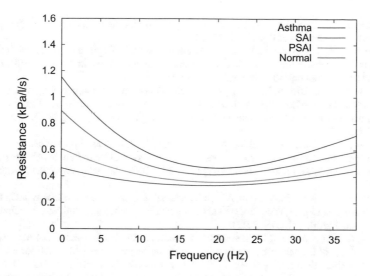

Fig. 12.4 Cubic resistance functions per class

- $\int |R_d(f) - (R_i^a(f) + \Delta R_{id})|\, df$, where $\Delta R_{id} \overset{\text{def}}{=} R_d(10) - R^a(10)$; and
- $\int |I_{R,d}(f) - (I_{R,i}^a(f) + \Delta I_{R,i,d})|\, df$, where $\Delta I_{R,i,d} \overset{\text{def}}{=} I_{R,d}(10) - I_{R,i}^a(10)$.

The resulting eight values serve as input to the neural network.

The results of training the neural network. The purpose of the neural network was to separate patients with lung dysfunctions from patients without lung dysfunction.

During the training, the network selected 50 neurons in the hidden layer. On the validation data set, the neural network achieved 100% accuracy on the validation set: all 74 cases were classified correctly.

This is much better that in the previous studies. The resulting classification accuracy is much better than the 60% accuracy achieved by neural networks without pre-processing.

References

1. Avila, N., Urenda, J., Gordillo, N., Kreinovich, V.: Scale-invariance-based pre-processing drastically improves neural network learning: case study of diagnosing lung dysfunction in children. University of Texas at El Paso, Department of Computer Science, Technical Report UTEP-CS-19-19 (2019)
2. World Health Organization (WHO), Asthma. http://www.who.int/mediacentre/factsheets/fs307/en/, downloaded on January 31, 2019
3. Mochizuki, H., Hirai, K., Tabata, H.: Forced oscillation techniques and childhood asthma. Allergol. Int. **61**(3), 373–383 (2012)

4. Bickel, S., Popler, J., Lesnick, B., Eid, N.: Impulse oscillometry: interpretation and practical applications. Chest **146**(3), 841–847 (2014)

5. Komarow, H.D., Myles, I.A., Uzzaman, A., Metcalfe, D.D.: Impulse oscillometry in the evaluation of diseases of the airways in children. Ann. Allergy Asthma Immunol. **106**(3), 191–199 (2011). https://doi.org/10.1016/j.anai.2010.11.011

6. Komarow, H.D., Skinner, J., Young, M., Gaskins, D., Nelson, C., Gergen, P.J., Metcalfe, D.D.: A study of the use of impulse oscillometry in the evaluation of children with asthma: analysis of lung parameters, order effect, and utility compared with spirometry. Pediatr. Pulmonol. **47**(1), 18–26 (2012)

7. Marotta, A., Klinnert, M.D., Price, M.R., Larsen, G.L., Liu, A.H.: Impulse oscillometry provides an effective measure of lung dysfunction in 4-year-old children at risk for persistent asthma. J. Allergy Clin. Immunol. **112**(2), 317–322 (2003). https://doi.org/10.1067/mai.2003.1627

8. Shi, Y., Aledia, A.S., Tatavoosian, A.V., Vijayalakshmi, S., Galant, S.P., George, S.C.: Relating small airways to asthma control by using impulse oscillometry in children. J. Allergy Clin. Immunol. **129**(3), 671–678 (2012)

9. Badnjevic, A., Cifrek, M.: Classification of asthma utilizing integrated software suite. In: Proceedings of the 6th European Conference of the International Federation for Mechanical and Biological Engineering IFMBE'2015, Dubrovnik, Croatia (2015), pp. 415–418

10. Badnjevic, A., Gurbeta, L., Cifrek, M., Marjanovic, D.: Classification of asthma using artificial neural networks. In: Proceedings of the 39th IEEE International Convention on Information and Communication Technology, Electronics, and Microelectronics MIPRO'2016, Opatija, Croatia (2016), pp. 387–390

11. Barua, M., Nazeran, H., Nava, P., Granda, V., Diong, B.: Classification of pulmonary diseases based on impulse oscillometric measurements of lung function using neural networks. In: Proceedings of the 26th Annual International Conference of the IEEE Engineering in Medicine and Biology Society, San Francisco, California, 1–5 Sept. 2004, pp. 3848–3851

12. Luo, G., Stone, B.L., Maloney, C.G., Gesteland, P.H., Yerram, S.R., Nkoy, F.L.: Predicting asthma control deterioration in children. BMC Med. Inf. Decis. Mak. **15**, Paper 84 (2015)

13. Sheskin, D.J.: Handbook of Parametric and Non-Parametric Statistical Procedures. Chapman & Hall/CRC, London, UK (2011)

14. Kreinovich, V., Lakeyev, A., Rohn, J., Kahl, P.: Computational Complexity and Feasibility of Data Processing and Interval Computations. Kluwer, Dordrecht (1998)

15. Vavasis, S.A.: Nonlinear Optimization: Complexity Issues. Oxford University Press, New York (1991)

16. Kearfott, R.B., Kreinovich, V.: Beyond convex? global optimization is feasible only for convex objective functions: a theorem. J. Glob. Optim. **33**(4), 617–624 (2005)

17. Huber, P.J., Ronchetti, E.M.: Robust Statistics. Wiley, Hoboken, New Jersey (2009)

18. Avila, N., Nazeran, H., Meraz, E., Gordillo, N., Aguilar, C.: Characterization of impulse oscillometric measures of respiratory small airway function in children. Adv. Electr. Electron. Eng. (2019). https://doi.org/10.15598/aeee.v17i1.2968

19. Nguyen, H.T., Kreinovich, V.: Applications of Continuous Mathematics to Computer Science. Kluwer, Dordrecht (1997)

20. Gholamy, A., Kreinovich, V., Kosheleva, O.: Why 70/30 or 80/20 relation between training and testing sets: a pedagogical explanation. Int. J. Intell. Technol. Appl. Stat. **11**(2), 105–111 (2018)

Chapter 13
Medical Application: Diagnostics, Part 2

Machine learning techniques have been very efficient in many applications, in particular, when learning to classify a given object to one of the given classes. Such classification problems are ubiquitous: e.g., in medicine, such a classification corresponds to diagnosing a disease, and the resulting tools help medical doctors come up with the correct diagnosis. There are many possible ways to set up the corresponding neural network (or another machine learning technique). A direct way is to design a single neural network with as many outputs as there are classes—so that for each class i, the system would generate a degree of confidence that the given object belongs to this class. Instead of designing a single neural network, we can follow a hierarchical approach corresponding to a natural hierarchy of classes: classes themselves can usually be grouped into a few natural groups, each group can be subdivided into subgroups, etc. So, we set up several networks: the first classifies the object into one of the groups, then another one classifies it into one of the subgroups, etc., until we finally get the desired class. From the computational viewpoint, this hierarchical scheme seems to be too complicated: why do it if we can use a direct approach? However, surprisingly, in many practical cases, the hierarchical approach works much better. In this chapter, we provide a possible explanation for this unexpected phenomenon.

Results from this chapter first appeared in [1].

13.1 Problem: Why Hierarchical Multiclass Classification Works Better Than Direct Classification

Case study: classification in medicine. One of the most challenging tasks in medicine is diagnostics. The main reason why this task is difficult is that usually, several different diseases have the same set of symptoms, and separating them is therefore not easy. To help medical doctors, researchers have been designing auto-

J. C. Urenda and V. Kreinovich, *Algebraic Approach to Data Processing*, Studies in Big Data 115, https://doi.org/10.1007/978-3-031-16780-5_13

matic software tools that provide a preliminary classification of patients into groups corresponding to different diseases. Of course, in their present form, these tools do not provide a perfect diagnostics, they cannot replace the medical doctors. However, these tools can be (and are) of help to medical doctors. They are especially useful for medical doctors who are just starting their medical careers and who therefore do not yet have the experience that comes from observing and treating hundreds and thousands of patients.

In particular, one of such tools has been developed by Nancy Avila for classifying children's lung dysfunctions [2].

How classification is usually obtained: general idea. One of the most widely used technique to get the desired classification is to use machine learning:

- we first train the algorithm on examples for which the classification is known;
- once the algorithm has been trained, we apply the resulting trained algorithm to new inputs and thus, produce the class that—according to this tool—contains this input.

At present, the most efficient machine learning tool is neural networks (see, e.g., [3, 4]), but other machine learning tools have been (and are) used to solve classification problems.

Multi-class classification: details. In many classification problems, we need to classify objects into several different classes. Let n denote the number of such classes.

In an ideal situation when, based on the evidence, it is absolutely clear that the object belongs to one of the classes, the system should produce this class. In many practical situations, however, even with all the available evidence, we are often not 100% sure what is the most appropriate class. In this case, we would like the system not only to produce one class, but to describe all possible classes—and for each possible class, to provide a degree of confidence that the object is in this class. This will definitely help the user to make a final decision.

In other words, to perform an n-class classification, we would like to design a system with n outputs $y_1, \ldots, y_n$, where each y_i describes to what extent the existing evidence supports the conclusion that the given object belongs to the i-th class. A user would also like to know the most probable class—i.e., in terms of the values y_i, the class i for which the degree of confidence y_i is the largest.

Direct approach. In the direct approach, we set up a neural network (or another machine learning tool) with n outputs, and we train it on the examples in which we know the class i_0, i.e., on the examples in which we have $y_{i_0} = 1$ and $y_i = 0$ for all $i \neq i_0$.

Hierarchical approach. Another alternative is to take into account that neural networks are, after all, attempts to simulate how we humans process data. Thus, when using neural networks, it makes sense to recall how we ourselves perform the corresponding analysis.

From this viewpoint, the direct approach is not how, e.g., medical doctors diagnose a patient—and, in general, not how we classify objects. Neither medical doctors not we humans in general keep in mind all thousands of possible alternatives. Instead, our classification is usually a hierarchical, multi-step one that goes from the general to the particular. For example, if we hear some not-very-loud noise when walking at night on a country road, we first check whether it is a human or an animal or something brought in by the wind. Then, if, e.g., it is an animal, we decide whether it is big (and possible dangerous) or small, if small whether it is a pet or, e.g., a rabbit, etc. The same logical process happens with doctors when diagnosing a disease, they start with general tests (i.e. blood testing) to identify normal or abnormal conditions; if an abnormal condition is observed, further testing (more specific) is performed to diagnose a disease.

In such cases, we reduce the large classification problem to several smaller-size ones, e.g., to several binary (2-class) classifications.

Empirical fact: hierarchical approach works much better. The direct approach is easier to implement: you just need to train one neural network. As a result, this is what people usually start with. Interestingly, this approach often does not work well, while the more difficult-to-implement hierarchical approach leads to much better learning and thus, to a much more accurate classification. This is, e.g., exactly what happened in the medical classification problem analyzed in [2].

How can we explain this empirical fact? In this chapter, we provide a possible explanation for this empirical fact.

13.2 Our Explanation

Analysis of the problem. We are interested in situations in which classification is difficult—otherwise, if it was easy, we would not need to design a complex computer-based tool to help the users.

In our scheme, we classify an object into a class i for which the degree of confident y_i (estimated by the system) is the largest. In these terms, *difficult* means that in many cases, it is difficult to decide which of the values y_i is the largest—i.e., that there are several classes i with approximately the same value of the degree of confidence.

For the "ideal" values $Y_1, \ldots, Y_n$ (that we would have gotten if we had all the information and no noise) we have $Y_{i_0} > Y_i$ for all $i \neq i_0$. However, because of the noise (and incompleteness of data) the estimates y_i are, in general, different from Y_i.

If the effect of the noise is that one of the values y_i with $i \neq i_0$ is "boosted", i.e., gets a reasonable-size positive estimation error $\Delta y_i \stackrel{\text{def}}{=} y_i - Y_i$, then the boosted value $y_i = Y_i + \Delta y_i$ becomes larger than $y_{i_0} \approx Y_{i_0}$ and thus, we get an incorrect classification.

To analyze whether a direct approach or a hierarchical approach work better, let us estimate, for each of these two approaches, the probability of such a misclassification.

What is the probability of a misclassification: case of the direct approach. Let p be a probability that one of the outputs gets boosted. Then, in the direct approach, we get a misclassification if one of the $n - 1$ wrong outputs gets boosted.

Each of these boostings is an independent event, so the probability that none of the $n - 1$ boosting occurs can be estimated as a product of $n - 1$ probabilities (equal to $1 - p$) that each of the boostings will not happen, i.e., as $(1 - p)^n$. Thus the probability that a misclassification will happen is equal to 1 minus this probability, i.e., to $1 - (1 - p)^{n-1}$.

The probability p is usually small—otherwise, the classification would be lousy. For small p, we can expand the above expression in Taylor series in p and ignore quadratic (and higher order) terms in this expansion. As a result, we conclude that the probability of misclassification if approximately equal to $(n - 1) \cdot p$.

What is the probability of a misclassification: case of the hierarchical approach. To make an estimation, let us consider the case when on each stage of the hierarchical classification, we have a binary classification—i.e., a classification into two classes. In this case, with one stage, we can classify objects into 2 classes; with 2 binary stages, we can classify them into 4 classes, and, in general, with s binary stages, we can classify objects into 2^s classes. Thus, to get a classification into n classes, we need to select the number of stages s for which $2^s \approx n$, i.e., we need $s \approx \log_2(n)$ stages.

On each stage, we compare two numbers y_i: the number corresponding to the correct group (or subgroup) and the number corresponding to an incorrect one. Here, it is also possible that boosting will resulting in a wrong subgroup. The probability of this, on each of the s stages, is p. If at one of the s stages, we get a wrong subgroup, we get a misclassification. Similarly to the direct case, we can conclude that the probability of misclassification is equal to $1 - (1 - p)^s \approx s \cdot p \approx \log_2(n) \cdot p$.

If we have 3 classes on each stage, we would get $2\log_3(n) \cdot p$. If we had 4 classes per stage, we would get $3\log_4(n) \cdot p$, etc.

The resulting explanation. In the direct approach, the probability of misclassification is $n \cdot p$, while in the hierarchical approach, this probability if $\log_2(n) \cdot p$. Since for large n, we have $\log_2(n) \ll n$, this shows that indeed the hierarchical classification is much more accurate—exactly as the empirical data shows.

This conclusion does not change if on each stage, we classify into $c \neq 2$ classes: we would get $(c - 1) \cdot \log_c(n) \cdot p$ which is still much smaller than $n \cdot p$.

References

1. Urenda, J., Avila, N., Gordillo, N., Kreinovich, V.: Hierarchical multiclass classification works better than direct classification: an explanation of the empirical fact. J. Uncertain Syst. **13**(3), 216–219 (2019)
2. Avila, N.: Computer-Aided Classification of Impulse Oscillometric Measures of Respiratory Small Airways Function in Children, Ph.D. Dissertation, Biomedical Engineering Program, University of Texas at El Paso (2019)

3. Bishop, C.M.: Pattern Recognition and Machine Learning. Springer, New York (2006)
4. Goodfellow, I., Bengio, Y., Courville, A.: Deep Learning. MIT Press, Cambridge, MA (2016)

Chapter 14
Medical Application: Diagnostics, Part 3

In many practical situations, the information comes not in terms of the original image or signal, but in terms of its Fourier transform. To detect complex features based on this information, it is often necessary to use machine learning. In the Fourier transform, usually, there are many components, and it is not easy to use all of them in machine learning. So, we need to select the most informative components. In this chapter, we provide general recommendations on how to select such components. We also show that these recommendations are in good accordance with two examples: the structure of the human color vision, and classification of lung dysfunction in children.

Results from this chapter first appeared in [1].

14.1 Problem: Which Fourier Components Are Most Informative

Fourier components are ubiquitous. In many practical situations, what we perceive or what we get from measuring instruments is not the original signal, but its Fourier transform. This is the case with our vision: we perceive the image in colors, i.e., by separating it into the corresponding Fourier components; see, e.g., [2, 3]. Similarly, when we hear music, we perceive it as sequence of notes, i.e., components corresponding to different frequencies.

This is true not only for our perceptions, this is also true for many measurement situations. For example, when a radio telescope observes a distant radio-source, the resulting signal is actually the Fourier transform of the original image; see, e.g., [4].

Need to select the most informative components. Images and signals are often difficult to process. In situations when we do not have exact formulas for detecting the desired features, it is often very efficient to apply machine learning techniques.

© The Author(s), under exclusive license to Springer Nature Switzerland AG 2022
J. C. Urenda and V. Kreinovich, *Algebraic Approach to Data Processing*,
Studies in Big Data 115, https://doi.org/10.1007/978-3-031-16780-5_14

However, it is difficult to directly apply these techniques to hundreds and thousands of data points that form each image or each signal. To be able to successfully apply these techniques, it is therefore desirable to select the most informative Fourier components.

14.2   Main Idea

First observation: magnitudes of Fourier components, in general, decrease with frequency. For a bounded-in-time signal or bounded-in-space image, the well-known Parceval Theorem states that the mean squared value of the image or signal $x(t)$ is equal to the mean square value of its Fourier transform $\widehat{x}(\omega)$: $\int x^2(t)\,dt = \int |\widehat{x}(\omega)|\,d\omega$. For a bounded signal or image, the integral $\int x^2(t)\,dt$ is finite, thus, the integral $\int |\widehat{x}(\omega)|\,d\omega$ is also finite. This implies that, on average, the absolute value $|\widehat{x}(\omega)|$ of the Fourier transform must decrease (and tend to 0) with frequency—otherwise, if this value did not decrease, the integral would be infinite.

Resulting first recommendation. The smaller the Fourier component, the less and less easy to distinguish it from the inevitable noise. Thus, the most informative component is the one which is the largest in (absolute) value—and thus, has the largest signal-to-noise ratio. Since, in general, the Fourier components decrease with frequency, a reasonable idea is to select the Fourier component $\widehat{x}(\omega)$ that corresponds to the smallest possible frequency ω_0.

In general, Fourier components are complex numbers $\widehat{x}(\omega) = r(\omega) + \mathrm{i} \cdot i(\omega)$, where $\mathrm{i} \overset{\text{def}}{=} \sqrt{-1}$. In these terms, the recommendation is to select the components $r(\omega_0)$ and $i(\omega_0)$.

What other components should we select: brainstorming. One complex-valued component may be not enough to detect the desired features. What other components should we select?

Let us recall that the real part $r(\omega)$ of the Fourier transform is proportional to $\int x(t) \cdot \cos(\omega \cdot t)\,dt$, and its imaginary part $i(\omega)$ is proportional to

$$\int x(t) \cdot \sin(\omega \cdot t)\,dt.$$

For localized signals located close to some value $t \approx t_0$, we thus get $r(\omega) \approx c \cdot \cos(\omega \cdot t_0)$ and $i(\omega) \approx c \cdot \sin(\omega \cdot t_0)$, for some constant c.

As we have already mentioned, the most informative components are the ones whose absolute value is the largest. Thus, for the real-valued components, the first two most informative components correspond to the values where the cosine is equal to ± 1, i.e., the values $\omega \approx 0$ and $\omega \approx \pi/t_0$. (And an even more informative is the difference between these two components, which is equal to $2c$.) Within this frequency range, from 0 to π/t_0, the most informative value of the imaginary part is when the sine is equal to 1, i.e., the value $\omega \approx 0.5 \cdot \pi/t_0$.

Thus, we arrive at the following general recommendation.

Resulting general recommendation. As the most informative Fourier components, we should take:

- the real and imaginary components $r(\omega_0)$ and $i(\omega_0)$ corresponding to the smallest possible frequency ω_0;
- one more real-valued component $r(\Omega)$ corresponding to some larger frequency Ω (that depends on the signal or image)—or, better yet, the difference $r(\omega_0) - r(\Omega)$, and
- an imaginary component $i\left(\dfrac{\omega_0 + \Omega}{2}\right)$ corresponding to a frequency which is exactly halfway between ω_0 and Ω.

What we will do now. We will show, on two examples, that this reasonable crude approximate recommendation actually leads to good results.

14.3 First Case Study: Human Color Vision

Discussion. This example is not about machine learning selecting the most informative features—it is about which most informative features biological evolution has selected.

What our recommendation suggests. Specifics of human vision is that in this case, it is difficult to separate real and imaginary parts of the signal. So, in this case, our recommendation means that we should select components corresponding to a smaller frequency ω, to a larger frequency Ω, and to a frequency which is exactly halfway between ω_0 and Ω.

Thus, to get the most informative understanding of images, we should select three equidistant frequencies:

$$\omega_0 < \frac{\omega_0 + \Omega}{2} < \Omega.$$

How human vision system actually works. The human vision system does select three different colors—i.e., three different frequencies [2, 3]:

- red, corresponding to 430–480 THz;
- green, corresponding to 540–580 THz; and
- blue, corresponding to 610–670 THz.

For 430–480 and 610–670, the midpoint is 520–575 THz, which is 4% close to the actual middle range of 540–580 THz.

But is this convincing? One may argue that, of course, the midpoint is somewhere in between the two frequencies, so no wonder it is close to the midpoint between them. By the same logic, one could get the same result if we considered, e.g., wavelength. Let us give it a try. In terms of wavelength:

- red corresponds to 635–700 nm,
- green corresponds to 520–560 nm, and
- blue corresponds to 450–490 nm.

Here, for 635–700 and 450–490, the midpoint is 442.5–595 which is only 6% close to the actual middle range of 520–560 nm. So, indeed, the frequency-based description—motivated by our arguments—is much closer to the actual human vision system.

14.4 Second Case Study: Classifying Lung Dysfunctions

Formulation of the problem. In this case study, we consider three types of lung dysfunctions: asthma, Small Airways Impairment (SAI), and Possible Small Airways Impairment (PSAI). To correctly classify lung dysfunction in children, a promising idea is to use Impulse Oscillometry System (IOS), where a periodic signal with 5 Hz is added to the airflow coming to the patient, and the resulting outflow is described by its Fourier components $r(f) + i \cdot i(f)$ corresponding to different frequencies f. Of course, since the signal is periodic, all the frequencies are proportional 5 Hz [5]. It turns out that the most informative frequencies are between 5 20 Hz. Which of the corresponding components should we choose?

What our recommendation says. According to our general recommendation, we should select:

- the components $r(5)$ and $i(5)$ corresponding to the smallest possible frequency 5 Hz,
- the component $r(f)$ corresponding to the largest of the most informative frequencies—in this case, 20 Hz (or, better yet, the difference between the corresponding components $r(5) - r(20)$), and
- the component $i(f)$ corresponding to the midpoint between 5 and 20 Hz.

In this case, the midpoint between 5 and 20 is 12.5 Hz. There is no component with exactly this frequency, but there are two closest frequencies (which are equally close to 12.5 Hz): 10 and 15 Hz.

Thus, according to our general recommendation, the most informative Fourier components should be $r(5), r(20)$ (or, better yet, $r(5) - r(20)$), $i(5), i(10)$, and $i(15)$.

Empirical data is in perfect accordance with our recommendation. To test our recommendation, we used the data collected by our colleague Erika Meraz; see [6] for details. This data contained data sets from 112 patients with known diagnoses.

For each of components of the Fourier transform, we tested how well this component can differentiate between two different diagnoses. Specifically, we evaluated the importance of each component by comparing the means of the two diagnoses to determine if they were statistically different (at the usual confidence level $p < 0.05$). We then selected the components that statistically significantly differentiated every pair of diagnoses—and these were exactly the components mentioned above: $r(5)$, $r(20)$ (or, better yet, the difference $r(5) - r(20)$), $i(5)$, $i(10)$, and $i(15)$; see [6] for details.

References

1. Urenda, J., Avila, N., Gordillo, N., Kreinovich, V.: Which Fourier components are most informative: general idea and case studies. J. Uncertain Syst. **13**(2), 138–141 (2019)
2. Hunt, R.W.G.: The Reproduction of Colour. Wiley, Chichester, UK (2007)
3. Wyszecki, G., Stiles, W.S.: Colour Science: Concepts and Methods. Quantitative Data and Formulae, Wiley, New York (2000)
4. Verschuur, G.L., Kellermann, K.I.: Galactic and Extra-Galactic Radio Astronomy. Springer, Berlin, Heidelberg, New York (2012)
5. Bickel, S., Popler, J., Lesnick, B., Eid, N.: Impulse oscillometry: interpretation and practical applications. Chest, vol. 146, No. 3, pp. 841–847 (2014)
6. Avila, N., Nazeran, H., Meraz, E., Gordillo, N., Aguilar, C.: Characterization of impulse oscillometric measures of respiratory small airway function in children. Advances in Electrical and Electronic Engineering (2019). https://doi.org/10.15598/aeee.v17i1.2968

Chapter 15
Medical Application: Treatment

In this chapter, we show that many aspects of complex biological processes related to wound healing can be explained in terms of the corresponding geometric symmetries.
   Results from this chapter first appeared in [1].

15.1   Problem: Geometric Aspects of Wound Healing

Geometric aspects of wound healing: a brief descrption. When a wound appears, the body starts a complex process of healing the wound; see, e.g., [2–4]. Among other important related processes, two processes affect the geometric shapes:

- within the first few minutes of injury, cells called platelets move to (and along) the wound boundary and change their shape to cover ("clot") this boundary—thus protecting the remaining part of the body; see, e.g., [5, 6];
- some time after that, epithelial (skin-forming) cells change their shape to elongated and slowly move in the direction orthogonal to the skin's boundary to cover the tissues exposed by the wound; see, e.g., [7].

A challenge. Wound healing is a very complex well-choreographed, and often, very successful process. However, in contrast to many other complex biological processes, it occurs very naturally, without a complex behavior coded into a DNA molecule, without a complex multi-neural nervous system controlling different stages. Most of this complex choreography occurs on the level of cells themselves. How can we explain this unexpectedly complex behavior of such a seemingly control-less system?

J. C. Urenda and V. Kreinovich, *Algebraic Approach to Data Processing*,
Studies in Big Data 115, https://doi.org/10.1007/978-3-031-16780-5_15

What we do in this chapter. In this chapter, we show that the geometric aspects of wound healing can be naturally explained by analyzing the corresponding local geometric symmetries.

15.2   What Are Natural Symmetries Here and What Are the Resulting Cell Shapes: Case of Undamaged Skin

Natural local symmetries of the undamaged skin surface. To understand the wound healing process, let us start with analyzing the natural local symmetries of the undamaged skin surface.

Locally, a skin surface is flat, so we can locally represent it as a plane. In general, a plane has the following geometric symmetries: rotations, shifts in two directions, and scaling $x \rightarrow \lambda \cdot x$. Overall, we have a 4-dimensional group of symmetries.

There are also discrete symmetries: reflections across a line, and mirror reflections—reflections across a point.

Out of the continuous symmetries, rotations are the only symmetries that can be *exact*, in the sense that it is quite possible to have a circular piece of skin which is perfectly invariant with respect to all rotations around its center. In contrast, shifts and scalings are only *approximate* symmetries: there is no way to select a piece of skin which would be invariant with respect to all shifts or with respect to all possible scalings: such an invariant region would have to coincide with the whole infinite plane.

Symmetry of skin cells: case of undamaged skin. A skin in not a homogeneous medium, it consists of cells. The cell shapes are usually somewhat complicated, but in the first approximation, we can describe their shapes in geometric terms.

We want to describe the shape of a cell boundary. In this description, we take into account that while the plane itself has a 4-dimensional symmetry group, the cell boundary cannot have all these symmetries: e.g., if it was shift-invariant, then with any point on this boundary, we would have all the points obtained from it by shift; thus, the cell boundary would take over the whole plane. Thus, to form a cell, some original symmetries must be broken.

Symmetry breaking is a process which is ubiquitous in real world and thus, well studied in physics. According to physics, the most probable are the transitions that break the smallest number of symmetries (and thus retain the most symmetries); see, e.g., [8, 9]. For example, when we heat a well-structured well-symmetric crystal, we do not directly get a fully amorphous gas stage; first, we get a liquid stage, in which some structure is retained.

The main symmetries here are rotations—since they are the only symmetries which are exact. The only shape which is invariant with respect to rotations is a circle. And skin sells indeed form a circle-type shape, with approximately the same size in all directions. (Of course, they are not exact circles, since skin cells cover the whole skin surface, and it is not possible to cover a planar area with non-intersecting circles.)

15.3 What If the Skin Is Damaged: Resulting Symmetries and Cell Shapes

What happens when there is a wound. A typical wound is a cut. Locally, a cut is a straight line. Thus, locally, instead of a plane, we have the shape of a half-plane, an area of the plane which is to one side of the cut line.

Cell shapes at different locations: discussion. At a location distant from the wound boundary, locally, the skin still has the same shape. So locally, we have the same symmetries, and thus, we expect that the cells retain the same shape.

However, for cells at the wound boundary, not all local symmetries hold: e.g., there is no longer invariance with respect to rotations. Since symmetries are different, we expect the shapes to be different as well. To find the resulting shapes, let us analyze what are the resulting symmetries.

What are natural symmetries of the damaged skin surface. As we have mentioned, the half-plane is no longer invariant with respect to rotations. Of all the shifts, it is only invariant with respect to shifts parallel to the cut. It is also still invariant with respect to scalings. Thus, instead of the original 4-dimensional symmetry group, we now only have a 2-dimensional symmetry group. In this case, both symmetries are approximate, since, as we mentioned, a finite region cannot be exactly shift-invariant or scale-invariant.

The new configuration also has some remaining discrete symmetries: namely, it is invariant with respect to a reflection across any line which is orthogonal to the wound boundary.

Resulting cell shapes. What are the cell shapes in the vicinity of the wound boundary? Similarly to the case of cell shapes for the undamaged skin, the typical cell shapes for the damaged skin should be invariant with respect to at least some of the original symmetries.

Let us first look for the shapes which are invariant with respect to both shifts and scaling. To simplify our analysis, let us take a coordinate system in which the wound boundary has the form $y = 0$, and the remaining skin is located in the half-plane $y > 0$. In these coordinates, if the cell boundary contains a point (x_0, y_0) with $y > 0$, then:

- due to shift-invariance, it will contain the points (x, y_0) for all possible values x, and thus,
- due to scale-invariance, it will contain all the points (x, y) with $y > 0$—i.e., it will coincide with the whole skin.

Since each cell is only a part of the skin, this means that for the maximally symmetric cell shapes, we cannot have points with $y > 0$. Thus, these cells must be located in the area $y = 0$—i.e., on the boundary of the wound.

Since we are considering cells on the boundary of the wound, the cell must contain at least one point on this boundary $y = 0$, i.e., a point $(x_0, 0)$ for some x_0. Due to shift-invariance, we can conclude that the cell must contain all the points $(x, 0)$ for different x—i.e., that the cell must fit the wound boundary.

What if we only have some symmetries? If we have shift-invariance, then we get the same shapes—elongated cells that cover the boundary. What if we only have scale-invariance? In this case, if we take, as the origin of the x-axis, the point where the cell intersects with the boundary, then this intersection point will have the coordinates $(0, 0)$. A cell is not a single point, it has to have some other points. If (x_0, y_0) is such a point, then, due to scale-invariance, it must also contain all the points $(\lambda \cdot x_0, \lambda \cdot y_0)$. In other words, the cell must contain all the points on a half-line (ray) that goes from $(0, 0)$ to this point (x_0, y_0). So, we have an elongated cell at some angle to the wound's boundary.

Most of such lines are only scale-invariant. However, some of these lines—namely, the ones which are orthogonal to the wound boundary—have an additional discrete symmetry: namely, the configuration including such line and the wound's boundary is invariant with respect to reflections across this line.

Conclusion: resulting shapes listed in the order of their relative frequency. According to our symmetry-based geometric analysis, the most frequent—and the first to appear—will be the maximally symmetric shapes, i.e., the elongated cells which are located directly on the boundary of the wound. As we have mentioned, this is exactly what we observed in the wound healing process: in the beginning, such cells indeed appear.

Next come cells which have the largest number of remaining symmetries—namely, the cells which are elongated and perpendicular to the wound boundary. This is also indeed what happens at the second stage of the wound healing.

So, the shapes of the emerging cells and the order in which these cells appear can indeed be explained by geometric symmetries.

Comment. After the above two major types of cells, we may have cells with fewer remaining symmetries—namely, elongated cells which are oriented at an angle to the wound boundary. Such cells have indeed been observed; see, e.g., [2–4].

15.4   Geometric Symmetries Also Explain Observed Cell Motions

For undamaged skin, cells do not move. In the undamaged skin, a cell is surrounded by other cells, so it has nowhere to move.

The situation is different on the wound boundary. When there is a wound, there is an open space to which cells can move. Let us use geometric symmetries to describe the most probable directions of the cells' movement.

How to find the most probable motion direction: main idea. To find the most probable direction of the cell's motion, it is reasonable to use the same idea when describing the cells' shapes: namely, we look for directions that maximally preserve the corresponding symmetries.

Case of maximally symmetric cells. The cells on the boundary are invariant with respect to both shift along the boundary and scaling. Thus, the line describing the direction of their most probable motion should also be invariant with respect to the same two symmetries.

One can easily see that the only such line is the line of the boundary itself. Thus, we conclude that these cells will most probably move along the wound's boundary.

Of course, this motion is only possible is there is space to move—i.e., if there is a part of the boundary which is not yet covered by such cells. So, the reshaping of the corresponding cells and their motions will continue until there is no more space to move—i.e., until the whole boundary is covered by such cells. This is exactly what is needed to form a cover for protecting the remaining part of the skin.

Case of second generation cells. Let us now consider the "second generation" cells, elongated cells whose direction is orthogonal to the wound's boundary. These cells are invariant with respect to scaling and reflection across their own line. Thus, the most probable line of their motion should be invariant with respect to the same symmetries.

One can see that the only such line is the line coinciding with the line of the cell itself. Thus, most probably, the cell will move in the same direction—orthogonal to the line of the boundary.

From the pure symmetry viewpoint, it can move in both directions: it can move towards the boundary, or it can move away from it. However, from the physical viewpoint, it cannot move away from the boundary—there the skin in intact, and there is no place to move. So, the only physically possible movement will be across the wound boundary. This way, such cells will spread into the wound area and cover it—this is exactly what we observe in the wound healing process.

Conclusion. So, a seemingly complex process of wound healing indeed requires no central control—its seemingly well-choreographed steps can be naturally explained by the corresponding geometric symmetries.

References

1. Urenda, J.C., Kreinovich, V.: "Geometric aspects of wound healing", University of Texas at El Paso, Department of Computer Science, Technical Report UTEP-CS-19-45
2. Clark, R.A.F. (ed.): The Molecular and Cellular Biology of Wound Repair. Kluwer Academic Publishers, Dordrecht (1996)

3. Gourdie, R.G., Myers, T.A. (eds.): Wound Regeneration and Repair: Methods and Protocols. Humana Press, New York (2013)
4. Stigler, B.: Algebraic model selection and identification using Groebner bases. Abstracts of the 2019 Southwest Local Algebra Meeting SLAM'2019, El Paso, Texas, 23–24 Feb. 2019
5. Rasche, H.: Haemostasis and thrombosis: an overview. Eur. Heart J. Suppl. **3**, Supplement Q, Q3–Q7 (2001)
6. Versteeg, H.H., Heemskerk, J.W., Levi, M., Reitsma, P.H.: New fundamentals in hemostasis. Physiol. Rev. **93**(1), 327–358 (2013)
7. Garg, H.G.: Scarless Wound Healing. Marcel Dekker, New York (2000)
8. Feynman, R., Leighton, R., Sands, M.: The Feynman Lectures on Physics. Addison Wesley, Boston, MS (2005)
9. Thorne, K.S., Blandford, R.D.: Modern Classical Physics: Optics, Fluids, Plasmas, Elasticity, Relativity, and Statistical Physics. Princeton University Press, Princeton, NJ (2017)

Chapter 16
Applications to Economics: How Do People Make Decisions, Part 1

In this section and in several following sections, we describe applications of algebraic techniques to economics. Economics is all about people making decisions. So how do people make decisions? There is a whole area of research called *decision theory* that describes how we make decisions—or, to be more precise, how a rational person should make decisions.

This theory is briefly overviews in this chapter. In the next three chapters, we explain the specifics of how people make decisions. In the remaining three economics-related chapters, we discuss problems related to economic stimuli and to investments.

Usually, this theory is applied to conscientious decisions, i.e., decisions that we make after some deliberations. However, it is reasonable to apply it also to decisions that we make on subconscious level—e.g., to decisions on what to remember and what not to remember: indeed, these decisions should also be made rationally.

Let us briefly recall the main ideas and formulas behind decision theory; for details, see, e.g., [1–4].

To make a reasonable decision, we need to know the person's preferences. To describe these preferences, decision theory uses the following notion of *utility*. Let us denote possible alternatives by $A_1, \ldots, A_n$. To describe our preference between alternatives in precise terms, let us select two extreme situations:

- a very good situation A_+ which is, according to the user, much better than any of the available alternatives A_i, and
- a very bad situation A_- which is, according to the user, much worse than any of the available alternatives A_i.

Then, for each real number p from the interval $[0, 1]$, we can form a lottery—that we will denote by $L(p)$—in which:

© The Author(s), under exclusive license to Springer Nature Switzerland AG 2022
J. C. Urenda and V. Kreinovich, *Algebraic Approach to Data Processing*,
Studies in Big Data 115, https://doi.org/10.1007/978-3-031-16780-5_16

- we get the very good situation A_+ with probability p and
- we get the very bad situation A_- with the remaining probability $1 - p$.

Clearly, the larger the probability p, the more chances that we will get the very good situation. So, if $p < p'$, then $L(p')$ is better than $L(p)$.

Let us first consider the extreme cases $p = 1$ and $p = 0$.

- When $p = 1$, the lottery $L(p) = L(1)$ coincides with the very good situation A_+ and is, thus, better than any of the alternatives A_i; we will denote this by $A_i < L(1)$.
- When $p = 0$, the lottery $L(p) = L(0)$ coincides with the very bad situation A_- and is, thus, worse than any of the alternatives A_i: $L(0) < A_i$.

For all other possible probability values $p \in (0, 1)$, for each i, the selection between the alternative A_i and the lottery $L(p)$ is not pre-determined: the decision maker will have to select between A_i and $L(p)$. As a result of this selection, we have:

- either $A_i < L(p)$,
- or $L(p) < A_i$,
- or the case when to the decision maker, the alternatives A_i and $L(p)$ are equivalent; we will denote this by $A_i \sim L(p)$.

Here:

- If $A_i < L(p)$ and $p < p'$, then $A_i < L(p')$.
- Similarly, if $L(p) < A_i$ and $p' < p$, then $L(p') < A_i$.

Based on these two properties, one can prove that for the probability $u_i \overset{\text{def}}{=} \sup\{p : L(p) < A_i\}$:

- we have $L(p) < A_i$ for all $p < u_i$ and
- we have $A_i < L(p)$ for all $p > u_i$.

This "threshold" value u_i is called the *utility* of the alternative A_i.

For every $\varepsilon > 0$, no matter how small it is, we have $L(u_i - \varepsilon) < A_i < L(u_i + \varepsilon)$. In this sense, we can say that the alternative A_i is equivalent to the lottery $L(u_i)$. We will denote this new notion of equivalence by $\equiv$: $A_i \equiv L(u_i)$. Because of this equivalence, if $u_i < u_j$, this means that $A_i < A_j$. So, we should always select an alternative with the largest possible value of utility.

This works well if we know exactly what alternative we will get. In practice, when we perform an action, we may end up in different situations—i.e., with different alternatives. For example, we may have alternatives of being wet without an umbrella and being dry with an extra weight of an umbrella to carry, but when we decide whether to take the umbrella or not, we do not know for sure whether it will rain or not, so we cannot get the exact alternative. In such situations, instead of knowing the exact alternative A_i, we usually know the probability p_i of encountering each alternative A_i when the corresponding action if performed. If we know several actions like thus, which action should we select?

Each alternative A_i is equivalent to a lottery $L(u_i)$ in which we get the very good alternative A_+ with probability u_i and the very bad alternative A_- with the remaining probability $1 - u_i$. Thus, the analyzed action is equivalent to a two-stage lottery in which:

- first, we select one of the alternatives A_i with probability p_i, and then
- depending on which alternative A_i we selected on the first stage, we select A_+ or A_- with probabilities, correspondingly, u_i and $1 - u_i$.

As a result of this two-stage lottery, we end up either with A_+ or with A_-. The probability of getting A_+ can be computed by using the formula of full probability, as $u = \sum_i p_i \cdot u_i$. So, the analyzed action is equivalent to getting A_+ with probability u and A_- with the remaining probability $1 - u$. By definition of utility, this means that the utility of the action is equal to u.

The above formula for u is exactly the formula for the expected value of the utility. Thus, we conclude that the utility of an action is equal to the expected value of the utility corresponding to this action.

References

1. Fishburn, P.C.: Utility Theory for Decision Making. Wiley, New York (1969)
2. Kreinovich, V.: Decision making under interval uncertainty (and beyond). In: Guo, P., Pedrycz, W. (eds.) Human-Centric Decision-Making Models for Social Sciences, pp. 163–193. Springer, Berlin (2014)
3. Luce, R.D., Raiffa, R.: Games and Decisions: Introduction and Critical Survey. Dover, New York (1989)
4. Raiffa, H.: Decision Analysis. Addison-Wesley, Reading (1970)

Chapter 17
Application to Economics: How Do People Make Decisions, Part 2

There are a lot of commonsense advices in decision making: e.g., we should consider multiple scenarios, we should consult experts, we should play down emotions. Many of these advices come supported by a surprisingly consistent quantitative evidence. In this chapter and in the following two chapters, on the example of the above advices, we provide a theoretical explanations for these quantitative facts.

Results these three chapters first appeared in [1]; the results of this chapter were first announced in [2].

17.1 Problem: Need to Consider Multiple Scenarios

How do companies make big decision and how often do they make right decisions? Analyzing dozens of cases, Nutt [3, 4] concluded that in the vast majority of cases, companies considered only one alternative.

It turns out that in such cases, the correct decision was made in half of the times (actually, slightly less than half); in other 50% of the cases, the decision led to a failure.

In several cases, companies considered two different alternatives before making a decision. In such cases, the companies were successful $\frac{2}{3}$ of the time.

How can we explain this empirical data?

© The Author(s), under exclusive license to Springer Nature Switzerland AG 2022
J. C. Urenda and V. Kreinovich, *Algebraic Approach to Data Processing*,
Studies in Big Data 115, https://doi.org/10.1007/978-3-031-16780-5_17

17.2 Our Explanation

Usually, a big company has one major competitor. Thus, a company's project leads to a success if this project is better than a project implemented by a competing company.

Let us first consider the case when a company considers only one alternative. Since the vast majority of companies only consider one alternative, it is reasonable to assume that the competitor also considers only one alternatives. One of the two considered alternatives is better. In our analysis, we consider both companies; so, the situation is symmetric: the probability that the first company's project is better is the same as the probability that the second company's project is better. These two probabilities should add up to 1, so each company prevails with probability 50%. Thus, the 50% observation is explained.

On the other hand, if a company consider two alternatives, then, since a competitor usually considers only one, now we have three possible projects to consider. The probability for each of these projects to be the best is the same—i.e., $\frac{1}{3}$. The first company wins if one of its two projects is the best—i.e.,

- either its first project is the best
- or its second project is the best.

The probability of this happening is equal to

$$\frac{1}{3} + \frac{1}{3} = \frac{2}{3}.$$

This explains the second empirical observation.

Comment. In the one-alternative case, we can also take into account that sometimes, the competitor considers two alternatives. In such cases, the probability for the first company to succeed is $\frac{1}{3}$. So:

- in most cases, the company succeeds with probability $\frac{1}{2}$, but
- in some cases, it succeeds with a lower probability $\frac{1}{3}$.

Thus, overall, the probability of success is slightly lower than $\frac{1}{2}$—which is exactly what was observed.

References

1. Urenda, J., Biney, F., Cardiel, M., De Lao, P., DesArmier, A., Dodson, N., Dodson, T., Gonzalez, S., Hinojos, L., Huerta, J., Jones, R., Martinez, O., Saldana Matamoros, C.A., Munoz, M., Kreinovich, V.: How to make decisions: consider multiple scenarios, consult experts, play down

emotions—quantitative explanation of commonsense ideas. J. Uncertain Syst. **13**(3), 220–223 (2019)

2. Biney, F., DesArmier, A., Dodson, N., Dodson, T., Saldaña Matamoros, C.A., Kreinovich, V.: How often do companies make right decisions: theoretical explanation of an empirical observation. In: Abstracts of the 23rd Joint UTEP/NMSU Workshop on Mathematics, Computer Science, and Computational Sciences, El Paso, Texas, 3 Nov. 2018

3. Johnson, S.: How to make a big decision. New York Times, p. SR10, 1 Sept. 2018

4. Nutt, P.C.: Why Decisions Fail: Avoiding the Blunders and Traps That Lead to Debacles. Berrett-Koehler publishers, San Francisco (2002)

Chapter 18
Application to Economics: How Do People Make Decisions, Part 3

18.1 Problem: Using Experts

It is known that the use of expert knowledge makes predictions more accurate. This makes perfect sense.

From the commonsense viewpoint, we can expect all kinds of improvements. Interestingly, it turns out that there is not much of variability: a typical improvement—as cited, e.g., in [1] on the example of meteorological temperature forecasts—is that the accuracy consistently improves by 10%.

How can we explain quantitative phenomenon?

18.2 Towards an Explanation

Use of expert knowledge means, in effect, that we combine an estimate produced by a computer model with an expert estimate.

Let σ_m and σ_e denote the standard deviations, correspondingly, of the model and of the expert estimate.

In effect, the only information that we have about comparing the two accuracies is that expert estimates are usually less accurate than model results: $\sigma_m < \sigma_e$. So, if we fix σ_e, then the only information that we have about the value σ_m is that it is somewhere between 0 and σ_e.

We have no reason to assume that some values from the interval $[0, \sigma_e]$ are more probable than others. Thus, it makes sense to assume that all these values are equally probable, i.e., that we have a uniform distribution on this interval; see, e.g., [2]. For this uniform distribution, the average value of σ_m is equal to $0.5 \cdot \sigma_e$. Thus, we have

$$\sigma_e = 2 \cdot \sigma_m.$$

© The Author(s), under exclusive license to Springer Nature Switzerland AG 2022
J. C. Urenda and V. Kreinovich, *Algebraic Approach to Data Processing*,
Studies in Big Data 115, https://doi.org/10.1007/978-3-031-16780-5_18

In general, if we combine two estimates x_m and x_e with accuracies σ_m and σ_e, then the combined estimate x_c—obtained by minimizing the sum

$$\frac{(x_m - x_c)^2}{\sigma_m^2} + \frac{(x_e - x_c)^2}{\sigma_e^2}$$

is

$$x_c = \frac{x_m \cdot \sigma_m^{-2} + x_e \cdot \sigma_e^{-2}}{\sigma_m^{-2} + \sigma_e^{-2}},$$

with accuracy

$$\sigma_c^2 = \frac{1}{\sigma_m^{-2} + \sigma_e^{-2}};$$

see, e.g., [3].

For $\sigma_e = 2\sigma_m$, we have $\sigma_e^{-2} = 0.25 \cdot \sigma_m^{-2}$, thus

$$\sigma_c^2 = \sigma_m^2 \cdot \frac{1}{1 + 0.25} = \sigma_m^2 \cdot \frac{1}{1.25} = 0.8 \cdot \sigma_m^2,$$

hence

$$\sigma_c \approx 0.9 \cdot \sigma_m.$$

So we indeed get a 10% increase in the resulting prediction.

Comment. This explanation was previously announced in [4].

References

1. Silver, N.: The Signal and the Noise: Why So Many Decisions Fail - But Some Don't. Penguin Press, New York (2012)
2. Jaynes, E.T., Bretthorst, G.L.: Probability Theory: The Logic of Science. Cambridge University Press, Cambridge, UK (2003)
3. Sheskin, D.J.: Handbook of Parametric and Non-Parametric Statistical Procedures. Chapman & Hall/CRC, London, UK (2011)
4. Urenda, J., Cardiel, M., Hinojos, L., Martinez, O., Kreinovich, V.: Expert knowledge makes predictions more accurate: theoretical explanation of an empirical observation. In: Abstracts of the 23rd Joint UTEP/NMSU Workshop on Mathematics, Computer Science, and Computational Sciences, El Paso, Texas, November 3 (2018)

Chapter 19
Application to Economics: How Do People Make Decisions, Part 4

19.1 Why Should We Play Down Emotions

There are very good people in this world, people who empathize with others, people who actively help others. Based on all the nice and helpful things that these good people do, one would expect that other people would appreciate them, cherish them, and that, in general, their attitude towards these good people would be positive. However, in real life, the attitude is often neutral or even negative. The resulting emotions hurt our ability to listen to their advice and thus, improve our decisions.

Why? Is there a rational explanation for these emotions?

19.2 Towards Explanation

Each person's happiness is determined not only by this person's satisfaction with life, but also by other people's happiness: it is difficult to enjoy good life if many people around you suffer. Let us denote the Person i's satisfaction with life by s_i, and this person's level of happiness by h_i. Then, h_i depends on s_i and on h_j for all other j.

In the first approximation, we can assume that this dependence is linear:

$$h_i = s_i + \sum_{j \neq i} a_{ij} \cdot h_j.$$

A very good person v is very happy when others are happy and suffers when others suffer, i.e., $a_{vj} \approx 1$ for all j.

Let us consider a simplified model in which everyone's satisfaction is the same $s_i = s > 0$, everyone's attitude to v is the same: $a_{jv} = a$, and we ignore attitude towards everyone else. Then, $h_v = s + n \cdot h_j$, where n is the number of people except

J. C. Urenda and V. Kreinovich, *Algebraic Approach to Data Processing*,
Studies in Big Data 115, https://doi.org/10.1007/978-3-031-16780-5_19

for v, and $h_j = s + a \cdot h_v$. Substituting the above expression for h_v into this formula, we get

$$h_j = s + a \cdot s + a \cdot n \cdot h_j,$$

so

$$h_j = \frac{a + a \cdot s}{1 - a \cdot n}.$$

If a is reasonably positive, i.e., if $a > \dfrac{1}{n}$, then $h_j < 0$—i.e., everyone will be unhappy.
Thus, the desire to be happy implies that $a < \dfrac{1}{n}$.

With n in billions, this explains why on average, the attitude should be either neutral or negative.

Commonsense explanation. From the common sense viewpoint, the above mathematics makes perfect sense: A very good person is unhappy if other people are unhappy. If we empathize with this person, we become unhappy too, and since people do not want to be unhappy, they prefer (at best) to ignore others' unhappiness—or even blame them for their own unhappiness.

Comment. The above theoretical explanation was previously announced in [1].

This explanation is somewhat similar to the explanation of other similar phenomena which was presented in [2] based on formal decision theory (see, e.g., [2–6]).

References

1. Jones, R., De La O, P., Gonzalez, S., Huerta, J., Munoz, M., Kreinovich, V.: Why attitude to good people is not always positive: explanation based on decision theory. In: Abstracts of the 23rd Joint UTEP/NMSU Workshop on Mathematics, Computer Science, and Computational Sciences, El Paso, Texas. Accessed 3 Nov 2018
2. Nguyen, H.T., Kosheleva, O., Kreinovich, V.: Decision making beyond Arrow's 'impossibility theorem', with the analysis of effects of collusion and mutual attraction. Int. J. Intell. Syst. **24**(1), 27–47 (2009)
3. Fishburn, P.C.: Utility Theory for Decision Making. Wiley, New York (1969)
4. Kreinovich, V.: Decision making under interval uncertainty (and beyond). In: Guo, P., Pedrycz, W. (eds.) Human-Centric Decision-Making Models for Social Sciences, pp. 163–193. Springer (2014)
5. Luce, R.D., Raiffa, R.: Games and Decisions: Introduction and Critical Survey. Dover, New York (1989)
6. Raiffa, H.: Decision Analysis. McGraw-Hill, Columbus (1997)

Chapter 20
Application to Economics: Stimuli, Part 1

How should we stimulate people to make them perform better? How should we stimulate students to make them study better? These are the questions that we study in this chapter and in the following chapter.

In this chapter, we deal with the fact that, as many experiments have shown, reward for good performance works better than punishment for mistakes. In this chapter, we provide a possible theoretical explanation for this empirical fact.

The results of this chapter first appeared in [1].

20.1 Problem: Why Rewards Work Better Than Punishment

Reward versus punishment: an important economic problem. One of the most important issue in economics is how to best stimulate people's productivity, what is the best combination of reward and punishment that makes people perform better.

This problem is ubiquitous. This problem rises not only in economics, it appears everywhere. How do we stimulate students to study better? How do we stimulate our own kids to behave better?

Empirical fact. In his famous book [2] summarizing his research, the Nobelist Daniel Kahneman—one of the fathers of behavioral economics—cites numerous researches that all confirm that reward for good performance, in general, works better than punishment for mistakes; see, e.g., p. 175.

© The Author(s), under exclusive license to Springer Nature Switzerland AG 2022
J. C. Urenda and V. Kreinovich, *Algebraic Approach to Data Processing*,
Studies in Big Data 115, https://doi.org/10.1007/978-3-031-16780-5_20

But why? Like many other facts from behavioral economics, this empirical fact does not have a convincing theoretical explanation.

What we do in this chapter. In this chapter, we provide a theoretical explanation for this empirical phenomenon.

20.2   Analysis of the Problem

What people want. People spend some efforts e, and, based on results of these efforts, they get some reward $r(e)$. In the first approximation, we can say that the overall gain is the reward minus the efforts, i.e., the difference

$$r(e) - e. \qquad (20.1)$$

A natural economic idea is that every person wants to maximize his/her gain, i.e., maximize the difference (20.1).

How to proceed. In view of the formula (20.1), to explain why rewards work better than punishments, we need to analyze what are the reward functions $r(e)$ corresponding to these two different reward strategies. Similarly to our derivation of the formula (20.1), in our analysis, we will use simplified "first approximation" models, i.e., models providing a good qualitative understanding of the situation.

What reward function corresponds to rewarding good performance. Crudely speaking, rewarding good performance means that:

- if the performance is not good, i.e., if the effort e is smaller than the smallest needed effort e_0, there is practically no reward:

$$r(e) = r_+ \qquad (20.1a)$$

for some $r_+ \approx 0$;
- on the other hand, the more effort the person uses, the larger the reward; in other words, every effort beyond e_0 is proportionally rewarded, i.e.,

$$r(e) = r_+ + c_+ \cdot (e - e_0), \qquad (20.2)$$

for some constant c_+.

The constant c_+ depends on the units used for measuring effort and reward: one unit of effort corresponds to c_+ units of reward. The two formulas:

- the formula (20.1a) corresponding to the case $e \leq e_0$, and
- the formula (20.2) corresponding to the case $e \geq e_0$,

can be combined into a single formula

$$r(e) = r_+ + \max(0, c_+ \cdot (e - e_0)) = r_+ + c_+ \cdot \max(0, e - e_0). \qquad (20.3)$$

This dependence has the following form:

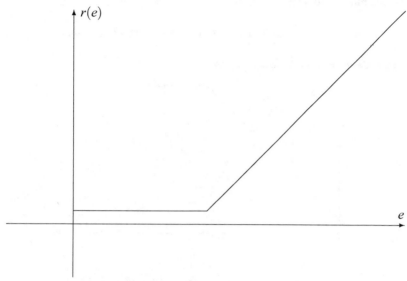

What can we say about this function. It is easy to see that the function (20.3) is *convex* in the sense that for all $e' < e''$ and for each $\alpha \in [0, 1]$, we have

$$r(\alpha \cdot e' + (1 - \alpha) \cdot e'') \leq \alpha \cdot r(e') + (1 - \alpha) \cdot r(e'').$$

What reward function corresponds to punishing for mistakes. Crudely speaking, punishing for mistakes means that:

- if the performance is good, i.e., if the effort e is larger than or equal to the smallest needed effort e_0, then there is no punishment, i.e., the reward remains the same:

$$r(e) = r_- \qquad (20.4)$$

for some constant r_-;
- on the other hand, the fewer effort the person uses, the most mistakes he/she makes, so the larger the punishment and the smaller the resulting reward; in other words, every effort below e_0 is proportionally penalized, i.e.,

$$r(e) = r_- - c_- \cdot (e_0 - e), \qquad (20.5)$$

for some constant c_-.

The constant c_- depends on the units used for measuring effort and reward: one unit of effort corresponds to c_- units of reward. The two formulas:

- the formula (20.4) corresponding to the case $e \geq e_0$, and
- the formula (20.5) corresponding to the case $e \leq e_0$,

can be combined into a single formula

$$r(e) = r_- - c_- \cdot \max(0, e_0 - e) = r_- + c_- \cdot \min(0, e - e_0). \qquad (20.6)$$

This dependence has the following form:

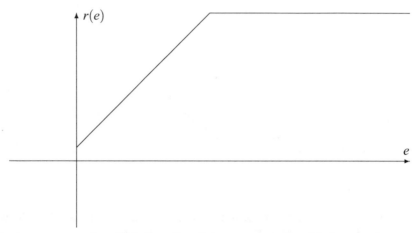

What can we say about this function. It is easy to see that this function is *concave* in the sense that for all $E' < E''$ and for each $\alpha \in [0, 1]$, we have

$$r(\alpha \cdot e' + (1 - \alpha) \cdot e'') \geq \alpha \cdot r(e') + (1 - \alpha) \cdot r(e'').$$

Now, we are ready to present the desired explanation.

20.3 Our Explanation

Known properties of convex and concave functions: reminder. It is known (see, e.g., [3]) that:

- every linear function is both convex and concave;
- the sum of two convex functions is convex, and
- the sum of two concave functions is concave.

In particular, the linear function $f(e) = -e$ is both convex and concave, thus:

- when the function $r(e)$ is convex, the sum $r(e) + (-e) = r(e) - e$ is also convex; and

- when the function $r(e)$ is concave, the sum $r(e) + (-e) = r(e) - e$ is also concave.

It is also known that:

- for a convex function, the maximum on an interval is always attained at one of this interval's endpoints, while
- for a concave function, its maximum on an interval is always attained at some point inside the interval.

Resulting explanation. As we have mentioned earlier, a person selects the effort e_0 for which the expression $r(e) - e$ attains its largest possible value.

Of course, people's abilities are not unbounded, there are certain limits within which we can apply the efforts. Thus, possible value of the effort e are located within some interval $[\underline{e}, \overline{e}]$. Thus:

- When we reward for good performance, the corresponding function $r(e)$ is convex, thus the difference $r(e) - e$ is convex, and therefore, the selected value e_0 coincides either with $\underline{e}$ or with $\overline{e}$. Thus, if we dismiss the case $e_0 = \underline{e}$ (when the reward is so small that it is not worth spending any effort), we conclude that $e_0 = \overline{e}$, i.e., the person selects the largest possible effort—which is exactly what we wanted to achieve.
- On the other hand, when we punish for mistakes, the corresponding function $r(e)$ is concave, thus the difference $r(e) - e$ is concave, and therefore, the selected value e_0 is always located inside the interval $[\underline{e}, \overline{e}]$: $e_0 < \overline{e}$. Thus, the person will not select the largest possible effort—which is exactly what we wanted to avoid.

This indeed explains why rewarding for good performance works better than punishment for mistakes.

Comments.

- What if we have both reward for good performance and punishment for mistakes, i.e., $r(e) = \text{const} + c_+ \cdot \max(0, e - e_0) + c_- \cdot \min(0, e - e_0)$? In this case, for $c_+ > c_-$, the function is still convex, i.e., we still get a very good performance, but if $c_- > c_+$, the function becomes concave, and the performance suffers. Thus, to get good results, reward must be larger than punishment.
- It is interesting to observe that the optimal rewarding function

$$r(e) = r_+ + c_+ \cdot \max(0, e - e_0),$$

in effect, coincides (modulo linear transformations of input and output) with the empirically efficient "rectified linear" activation function $r(e) = \max(0, e)$ used in deep learning; see, e.g., [4–6]. So, not only people learn better when we use this function—computers learn better too!

References

1. Kosheleva, O., Urenda, J., Kreinovich, V.: Reward for good performance works better than punishment for mistakes: economic explanation, University of Texas at El Paso, Department of Computer Science, Technical Report UTEP-CS-20-40 (2020)
2. Kahneman, D.: Thinking, Fast and Slow. Farrar, Straus, and Giroux (2011)
3. Rockafeller, R.T.: Convex Analysis. Princeton University Press, Princeton, New Jersey (1970)
4. Fuentes, O., Parra, J., Anthony, E., Kreinovich, V.: Why rectified linear neurons are efficient: a possible theoretical explanations. In: Kosheleva, O., Shary, S., Xiang, G., Zapatrin, R. (eds.) Beyond Traditional Probabilistic Data Processing Techniques: Interval, Fuzzy, etc, pp. 603–613. Methods and Their Applications, Springer, Cham, Switzerland (2020)
5. Goodfellow, I., Bengio, Y., Courville, A.: Deep Learning. MIT Press, Cambridge, Massachusetts (2016)
6. Kreinovich, V., Kosheleva, O.: Deep learning (partly) demystified. In: Proceedings of the 2020 4th International Conference on Intelligent Systems, Metaheuristics & Swarm Intelligence ISMSI'2020. Thimpu, Bhutan, 18–19 April 2020

Chapter 21
Application to Economics: Stimuli, Part 2

In this chapter, we deal with perceived fairness. At first glance, it seems that people should be paid in proportion to their contribution, so if one person produces a little more than the other one, he/she should be paid a little more. In reality, however, top performers are paid dis-proportionally more than those whose performance is slightly worse. How can we explain this from an economic viewpoint? We show that actually there is no paradox here: a simple economic analysis shows that in many area, it makes perfect economic sense to pay much more to top performers.

Results from this chapter first appeared in [1].

21.1 Problem: Why Top Experts Are Paid So Much

Top experts are well paid. Whatever area we take, top experts are paid much more than those who are almost on the same level:

- top athletes get multi-million dollar contracts while those who can run, swim, etc., only slightly worse, get paid (if at all) several orders of magnitude less;
- top managers get millions of dollars, while managers who seem to have almost similar skills—but somewhat worse success rate—get paid much less: the difference between the salaries of the highest paid manager and the next highest is usually huge.

The same phenomenon occurs in many areas of activity such as book publishing, movie making, etc.; see, e.g., [2–4] and references therein. Even university professors—although their salaries are much more equal—follow this trend: the salary of the highest-paid professor is more than an order of magnitude higher than the salary of the lowest-paid US professor.

© The Author(s), under exclusive license to Springer Nature Switzerland AG 2022
J. C. Urenda and V. Kreinovich, *Algebraic Approach to Data Processing*,
Studies in Big Data 115, https://doi.org/10.1007/978-3-031-16780-5_21

From the economic viewpoint, this seem to be paradoxical. At first glance, from the economic viewpoint, this seems to be a paradox: in economics, everyone's pay should be proportional to this person's contributions, so why should a small difference in performance lead to such a huge difference in salary?

If a company pays $100 K a year to a highly qualified worker, and then an even more qualified worked who is 10% better applies for the job, a reasonable ideas seems to pay this new person 10% more, i.e., $110 K per year—but not 10 times more. So why such a seeming overpayment of top professionals consistently happens in businesses where economy should be the main driving force? Even if we discard public universities which have other criteria of success, there are plenty of other examples where top professionals are seemingly overpaid. How can we explain this?

What we do in this chapter. In this chapter, we show that very high salaries of top experts actually make economic sense. Specifically, we provide a simplified model of this phenomenon—simplified enough so that it can be easily analytically studied—and we show that already in this simplified model, reasonable behavior leads to exactly the "overpayment" phenomenon—that top experts who are even slightly better get paid much more than their nearest competitors.

21.2   Our Explanation

How this phenomenon can be explained in the idealized case. As an example, let us consider investment fund managers. The quality of a money manager is determined by the return on investment that this person can achieve: better managers invest smarter and thus, provide a better return on investment, while not so good managers provide smaller return on investment.

In real life, returns on investment vary from one year to another. So, when we talk about quality of money managers, we need to take into account their average return on investment over a certain period of time.

The fund usually gets, every year, a certain percentage p of the invested money. Let us consider an ideal situation, in which every potential investor known the average return on investment r_i of each money manager i. Then, for each dollar invested with the ith manager, the investor will get, on average, the additional amount $r_i - p$.

Each investor wants to maximize his/her amount of money. So, each investor will invest in a fund with the largest possible value of $r_i - p$. So, in this idealized situation, everyone will invest in the fund whose money manager is the superstar— i.e., the fund i_0 for which the value r_i is the largest: $r_{i_0} = \max_i r_i$. This fund will earn money from all these investments, while all other funds will have no money to manage at all and thus, will not survive. A money manager for which r_i is almost the same as r_{i_0} but slightly smaller will earn nothing, while the find that hires the superstar money manager will earn billions. Because of this difference, it pays to provide a huge salary to the superstar manager.

Similarly, a top engineer who come up with a slightly better and/or slightly cheaper design will help the company take over the whole market for the corresponding gadgets—while others, whose gadgets are slightly worse or slightly more expensive, will not survive.

What happens in more realistic situations. In reality, e.g., for money managers, their rates of return vary so much that it is difficult to accurately estimate the average rate of return. We can compute the arithmetic average of the past rates of return, but, as is well known, for small samples, the sample average is somewhat different from the expected value; see, e.g., [5]. In addition to arithmetic average, there are other possible statistics that estimate the expected value—e.g., for symmetric distributions, we can take the median, or we can take some robust method; see, e.g., [6].

In such case, all we can do is select a manager with the largest value of the estimated return. For a finite sample, for which the sample-based estimates differ from the actual expected value, based on different estimates, we may select different managers as the best. Thus, the managers who are slightly worse than the best one do have a chance to be selected – so the situation is not that catastrophic for them as in the idealized situation. However, the more accurate our estimates, the smaller the chance that they will be selected – and so, the smaller the average salary of such managers.

References

1. Urenda, J., Kreinovich, V.: Why top experts are paid so much: economics-based explanation. Appl. Math. Sci. **13**(12), 591–594 (2019)
2. Adler, M.: Stardom and talent. Am. Econ. Rev. **75**(1), 208–212 (1985)
3. Barabasi, A.-L.: The Formula: The Universal Laws of Success. Brown, and Company, New York, Little (2018)
4. Rosen, S.: The economics of superstars. Am. Econ. Rev. **71**(5), 845–858 (1981)
5. Sheskin, D.J.: Handbook of Parametric and Non-parametric Statistical Procedures. Chapman & Hall/CRC, London, UK (2011)
6. Huber, P.J., Ronchetti, E.M.: Robust Statistics. Wiley, Hoboken, New Jersey (2009)

Chapter 22
Application to Economics: Investment

When people have several possible investment instruments, people often invest equally into these instruments: in the case of n instruments, they invest $1/n$ of their money into each of these instruments. Of course, if additional information about each instrument is available, this $1/n$ investment strategy is not optimal. We show, however, that in the absence of reliable information, $1/n$ investment is indeed the best strategy.

Results of this chapter first appeared in [1].

22.1 $1/n$ Investment: Formulation of the Problem

General investment problem. People saving for retirement usually have several options to invest: they can invest in stocks, they can invest in bonds, they can invest in funds that combine stocks and bonds, etc. An important decision is how to allocate money between different financial instruments, i.e., how much money should we invest in each of the instruments.

Markowitz's solution. A solution to this problem was proposed in 1952 by the future Nobelist Markowitz [2]. He actually solved two different versions of the investment problem:

- the first version is when we want to achieve a certain expected growth rate, and within this expected rate, we select an investment portfolio that minimizes the risk (as measured by the standard deviation of the growth rate);
- the second version is when we fix the risk level, and within this risk level, we select an investment portfolio that maximizes the expected growth rate.

© The Author(s), under exclusive license to Springer Nature Switzerland AG 2022   113
J. C. Urenda and V. Kreinovich, *Algebraic Approach to Data Processing*,
Studies in Big Data 115, https://doi.org/10.1007/978-3-031-16780-5_22

What is $1/n$ **investment**. In the absence of reliable information, people tend to divide their investment amount equally between different investment options: if there are n investment options, they invest exactly $1/n$ of the original amount in each of these options. This practice is known as the $1/n$ *investment*; see, e.g., [3, 4].

Numerous experiments confirm that this is how, in the absence of detailed information, people invest their money [3–7]. We are not talking only about common folks who do not understand the corresponding economic details: this is, e.g., how Markowitz himself invested his retirement money [4, 8].

Why? A natural question is: why is this strategy so ubiquitous? This question is formulated, e.g., in [4]—this book tends to use the ubiquity of this strategy as one of the arguments that people do not always behave rationally.

In this chapter, we provide a natural and simple explanation for this strategy—which shows that at least in this case, people do behave rationally.

22.2   Our Explanation

Main idea. The absence of information about the available investment options means that we have no reason to assume that one of them is better than the other. In other words, based on the available information (or, to be more precise, based on the absence of any information), the situation is completely invariant with respect to any permutation of the available n options.

It is therefore reasonable to conclude that the resulting allocation of the available money amount between different investment options should also be invariant with respect to the same transformations.

This idea explains the $1/n$ **investment strategy**. Allocating funds means selecting, for each investment option i, the proportion $p_i \geq 0$ that will be invested in this particular option. The only a priori restriction is that all the money should be invested, i.e., that these proportions should all up to 1: $p_1 + \ldots + p_n = 1$.

Our requirement is that the allocation should be invariant with respect to any permutation $\pi : \{1, \ldots, n\} \to \{1, \ldots, n\}$. This means, in particular, that for every i and j, the allocation $p_1, \ldots, p_n$ should be invariant with respect to the permutation that swaps i and j and leaves all other elements unchanged. This invariance means that we should have $p_i = p_j$. Since this should be true for all i and j, all n allocations $p_1, \ldots, p_n$ should be equal to each other. Since their sum should be equal to 1, this means that each of them should be equal to $1/n$—thus, in the absence of information, the $1/n$ investment is indeed the most reasonable strategy.

22.3 Discussion

What about probabilistic investment strategies? In the previous section, we considered only *deterministic* investment strategies, in which we select the allocations p_i. However, in principle, we can have *probabilistic* investment strategies, in which, instead of selecting an allocation vector $p = (p_1, \ldots, p_n)$ from the very beginning, we could:

- select a probability distribution on the set of all allocation vectors p, and then
- when it comes to each real investment opportunity, select one of the vectors p with the corresponding probability.

Why probabilistic strategies make sense. At first glance, this may seem like a purely mathematical exercise, but the experience of game theory has shown that in many situation, such a probabilistic choice works much better than any deterministic one; see, e.g., [9].

This fact is easy to explain. As an example, let us consider a simpler robbers-and-cop situation, when robbers want to rob one of the two banks, and a local police department only has enough folks to protect a single bank. In this case, if the police department selects a deterministic strategy—i.e., allocated all its resources into one of the banks all the time—this will be a disaster: robbers will know which bank is protected and thus rob the other bank. The only strategy avoiding such a disaster is a probabilistic one, in which each time, the police officers are allocated to each bank with probability 1/2. Then, the robbers have a 50% chance of being caught no matter which bank they select, and this will most probably deter them from attempting an attack.

Probabilistic strategies reduce to deterministic ones. Suppose that we have a probabilistic strategy for investment. This strategy is optimal for any possible investment value.

Suppose now that we have an amount x that we want to invest. One possibility is to use the optimal probabilistic strategy to invest the whole amount. However, alternatively, we can divide the original amount x into two smaller amounts $x/2$ each. For each of these two amounts, we can use the same optimal probabilistic investment strategy. In this case, the overall investment is equal to the sum of two independent identically distributed random variables.

Instead of dividing the investment amount into two equal parts, we can similarly divide it into N equal parts, for each integer n. In this case, the resulting investment is the sum of N independent identically distributed random variables.

For large N, the Law of Large Numbers applies (see, e.g., [10]), so the resulting distribution will simply concentrate on the expected allocation for each of the instruments—i.e., in this case, due to symmetry, on the allocation in which we assign the same proportion $1/n$ to each of these instruments. Thus, if a probabilistic strategy is optimal, the corresponding deterministic $1/n$ investment strategy is also optimal.

References

1. Urenda, J., Kreinovich, V.: In the absence of information, $1/n$ investment makes perfect sense. Appl. Math. Sci. **13**(12), 585–589 (2019)
2. Markowitz, H.M.: Portfolio selection. J. Finance **7**(1), 77–91 (1952)
3. Benartzi, S., Thaler, R.H.: Naive diversification strategies in defined contribution saving plans. Am. Econ. Rev. **91**(1), 79–98 (2001)
4. Thaler, R.H., Sunstein, S.R.: Nudge: Improving Decisions about Health, Wealth, and Happiness. Penguin Books, New York (2009)
5. Benartzi, S., Thaler, R.H.: Heuristics and biases in retirement savings behavior. J. Econ. Perspect. **21**(3), 81–104 (2007)
6. Read, D., Loewenstein, G.: Diversification bias: explaining the discrepancy in variety seeking between combined and separated choices. J. Exp. Psychol. Appl. **1**, 34–49 (1995)
7. Simonson, I.: The effect of purchase quantity and timing on variety-seeking behavior. J. Mark. Res. **28**, 150–162 (1990)
8. Zweig, J.: Five investing lessons from America's top pension fund. Money 115–118 (1998)
9. Maschler, M., Solan, E., Zamir, S.: Game Theory. Cambridge University Press, Cambridge, UK (2013)
10. Sheskin, D.J.: Handbook of Parametric and Non-parametric Statistical Procedures. Chapman & Hall/CRC, London, UK (2011)

Chapter 23
Application to Social Sciences: When Revolutions Happen

In this chapter, we provide an application to social sciences, namely, to predicting the most explosive social event—a revolution.

At first glance, it may seem that revolutions happen when life becomes really intolerable. However, historical analysis shows a different story: that revolutions happen not when life becomes intolerable, but when a reasonably prosperous level of living suddenly worsens. This empirical observation seems to contradict traditional decision theory ideas, according to which, in general, people's happiness monotonically depends on their level of living. A more detailed model of human behavior, however, takes into account not only the current level of living, but also future expectations. In this chapter, we show that if we properly take these future expectations into account, then we get a natural explanation of the revolution phenomenon.

Results from this chapter first appeared in [1].

23.1 Formulation of the Problem

When revolutions happen: usual understanding. People usually believe that revolutions happen when the situation worsens to such extent that life under the old regime becomes practically intolerable. Paraphrasing the famous saying attributed to Marie-Antoinette, people start a revolution when they do not even have enough bread to eat.

When revolutions actually happen. However, a historical analysis shows that the usual understanding is wrong; see, e.g., [2–4]. Most revolutions happen *not* when the situation is at its worst, they usually happen when the situation has been improving for some time and then suddenly gets worse—although, by the way, never as bad as it was before the improvement started.

How can we explain this? This is an interesting observation, but it leaves one puzzled: why? There are well-designed theories of human decision making, and

© The Author(s), under exclusive license to Springer Nature Switzerland AG 2022
J. C. Urenda and V. Kreinovich, *Algebraic Approach to Data Processing*,
Studies in Big Data 115, https://doi.org/10.1007/978-3-031-16780-5_23

experiments show that in most situations, people act rationally: the more their needs are satisfied, in general, the happier they are.

So how come that right before the revolution, when the level of living is higher (often much higher) than in the recent past, people are so much less happy that they start a revolution—while in the past, when their living conditions were much worse, they were sufficiently satisfied—at least so as to remain obedient.

How can we explain this unexpected (and somewhat counterintuitive) behavior?

What we do in this chapter. In this chapter, we show that this seemingly counterintuitive revolution phenomenon can actually be well explained within the standard decision theory.

23.2   Analysis of the Problem

Traditional decision theory: a brief reminder. According to traditional decision theory, people's preferences are described by numerical values called *utilities*; see, e.g., [5–9].

The actions of a person are determined not just by this person's current level of satisfaction—as described by the current utility value u_0—but also by the expected future utility values u_1 at the next moment of time, u_2 in the second next moment of time, etc. The future utility values come with a discount; e.g., the possibility to buy a new car a few years in the future does not bring as much happiness as buying a car right away.

This is similar to the value of future money: future money is less valuable than the same amount right now, since if we have the same amount—say $1000—now, we can place it in a bank and, due to accumulating interest, get a larger amount in the future. If we denote the annual interest rate by α, then after year t, each invested dollar will turn into $(1 + \alpha)^t$ dollars. Thus, $1 at time t is equivalent to q^t dollars now, where we denoted $q \stackrel{\text{def}}{=} \dfrac{1}{1+\alpha}$. So, if we get the amount a_0 now, the amount a_1 in the next year, the amount a_2 in 2 years, etc., this is equivalent to getting the following amount now:

$$a_0 + q \cdot a_1 + q^2 \cdot a_2 + \cdots$$

A similar formula can be used to describe the overall utility based on the current utility u_0 and expected future utilities

$$u_0 + q \cdot u_1 + q^2 \cdot u_2 + \cdots$$

This general approach requires extrapolation. The future amounts are based on extrapolation. So, to apply this theory to our situation, we need to understand how exactly people extrapolate.

In general, extrapolations means that:

- we select a family of functions characterized by a few parameters

$$u_t = f(p_1, \ldots, p_n, t),$$

- then we find the values $\widehat{p}_1, \ldots, \widehat{p}_n$ of the parameters that best fit the observed data u_0, u_{-1}, u_{-2}, etc., i.e., for which

$$f(p_1, \ldots, p_n, 0) \approx u_0, \quad f(p_1, \ldots, p_n, -1) \approx u_{-1}, \ldots,$$

- and then we use these values to predict future values as $f(\widehat{p}_1, \ldots, \widehat{p}_n, t)$.

It is reasonable to use models which are linear in the its parameters. A reasonable idea is to use models that linearly depend on the corresponding parameters: for such models, matching parameters to data means solving systems of linear equations, which is very feasible and much easier than solving systems of nonlinear equations—which are, in general, NP-hard.

Thus, we consider models of the type $u_t = \sum_{i=1}^n p_i \cdot f_i(t)$, where $f_i(t)$ are given functions, and p_i are appropriate parameters.

Which basis functions $f_i(t)$ should we choose? Most transitions are smooth, so it is reasonable to require that all the functions $f_i(t)$ used to extrapolation are smooth.

Another reasonable requirement is related to the fact that the numerical value of time depends on the choice of a measuring unit—years or months—and on the choice of a starting time. For example, during the French revolution, the year of storming the Bastille was considered Year 1.

If we change a measuring unit by a new one which is a times smaller, then each original value t is replaced by the new value $a \cdot t$. Similarly, if we change the original starting point with the new starting point which is b units in the past, then the original value t is replaced by the new value $t + b$.

The general formulas for extrapolation should not depend on such an arbitrary things as selecting a unit of time or selecting a starting point. It is therefore reasonable to assume that the approximating family $\left\{ \sum_{i=1}^n p_i \cdot f_i(t) \right\}$ will not change if we simply re-scale time to $t \to a \cdot t$ or to $t \to t + b$.

In other words, we require that for every $a > 0$ and for every b, we have

$$\left\{ \sum_{i=1}^n p_i \cdot f_i(a \cdot t) \right\}_{p_1, \ldots, p_n} = \left\{ \sum_{i=1}^n p_i \cdot f_i(t + b) \right\}_{p_1, \ldots, p_n} = \left\{ \sum_{i=1}^n p_i \cdot f_i(t) \right\}_{p_1, \ldots, p_n}.$$

It turns out that under these conditions, all the basic functions—and thus, all their linear combinations—are polynomials; see, e.g., [10].

Thus, it is reasonable to approximate the actual history by a polynomial. Let us show, on a simple example, that this indeed explains the empirical revolution phenomenon.

Two simple situations. Specifically, we will compare two simple situations:

- a situation in which the level of living is consistently bad, i.e.,

$$u_0 = u_{-1} = \cdots = u_{-k} = \cdots = c_1$$

for some small value c_1, and
- a situation in which the level of living used to be much better, but now somewhat decreased, i.e., in which

$$u_{-1} = u_{-2} = \cdots = c_+$$

but $u_0 = c_- < c_+$—although this decreased value $u_0 = c_-$ is still better than the value c_1 from the first situation.

If people did not take their future happiness into account when making decision, the situation would have been very straightforward—and in full accordance with the commonsense understanding of the revolutions: people in the first situation would be much less happy than people in the second situation and therefore, more prone to start a revolution.

What will happen if we take future expectations into account? In the first situation, of course, a reasonable extrapolation should lead to the exact same small value $u_0 = c$; thus, the overall utility is equal to

$$u_0 + q \cdot u_1 + \cdots = c \cdot (1 + q + q^2 + \cdots) = \frac{c}{1 - q}.$$

But what to expect in the second situation?

Let us start with the simplest possible extrapolation. Let us start our analysis with the simplest possible extrapolation, when we make our future predictions based only on two utility values: the current utility value u_0 and the previous utility value u_{-1}.

Which degree polynomials should we use? In this case, we have two values u_0 and u_{-1} to fit the model, so it is reasonable to select the degree of the approximating polynomial for which the corresponding family of polynomials depends on exactly two parameters. Polynomials of a general degree d have the form

$$a_0 + a_1 \cdot t + \cdots + a_d \cdot t^d.$$

This family depend on $d + 1$ parameters a_i, so in our case, we should have $d + 1 = 2$ and $d = 1$—i.e., we should use linear functions for extrapolation.

Since $u_0 < u_{-1}$, we thus get a linear decreasing function. Its values tend to $-\infty$ as the time t increases. So, when q is close to 1, the corresponding value

$$u_0 + q \cdot u_1 + \cdots \approx u_0 + u_1 + u_2 + \cdots$$

becomes very negative—and this explains why in the second situation, the revolution is much more probable.

What about more realistic approximation schemes? One may think that the above explanation is caused by our oversimplification of the extrapolation model. Of course, linear extrapolation is a very crude and oversimplified idea. What happens if we use higher degree polynomials for extrapolation?

Let us assume that for extrapolation, we use polynomials of order d. The corresponding family of polynomials have d_1 parameters, so we can fit $d + 1$ values. Thus, if we use these polynomials, then, in our extrapolation, we can use not only the two values u_0 and u_{-1}, we can use $d + 1$ values

$$u_0, u_{-1}, \ldots, u_{-d}.$$

Let us find the polynomial $P(t)$ of degree d that fits all these values, i.e., for which $P(-i) = c_+$ for all i from 1 to d, and $P(0) = c_-$. These conditions become even easier if we consider an auxiliary polynomial $Q(t) \overset{\text{def}}{=} P(t) - c_+$. For this auxiliary polynomial, we have $Q(-d) = \ldots = Q(-1) = 0$ and $Q(0) = c_- - c_+$. This polynomial of degree d has d roots $t = -1$, ..., $t = -d$, thus, it is divisible by the monomials $t - (-i) = t + i$ for all i from 1 to d, and therefore, it has the form $Q(t) = C \cdot (t + 1) \cdot (t + 2) \cdot \ldots \cdot (t + d)$, for some constant C. This constant can be determined from the condition that $Q(0) = c_- - c_+$, so $C \cdot 1 \cdot 2 \cdot \ldots \cdot d = c_- - c_+$ and thus,

$$C = \frac{c_- - c_+}{1 \cdot 2 \cdot \ldots \cdot d}.$$

Therefore, for any $t > 0$, the extrapolated value of $P(t) = c_+ + Q(t)$ has the form

$$Q(t) = c_+ + (c_- - c_+) \cdot \frac{(t + 1) \cdot (t + 2) \cdot \ldots \cdot (t + d)}{1 \cdot 2 \cdot \ldots \cdot d}.$$

Since $c_- < c_+$, this value is negative—and tends to $-\infty$ as the time t increases. In comparison with the linear extrapolation case, it tends to $-\infty$ even faster than in the case of linear extrapolation—as t^d.

So, *the revolution phenomenon can be explained* no matter what degree of extrapolation we use.

Discussion. Based on our analysis, in addition to our main conclusion (that we have explained the seemingly counterintuitive revolution phenomenon), we can make two auxiliary conclusions (which also fit perfectly well with common sense):

- revolutions only happen if people care about the future; if they don't, if $q \approx 0$, people are happy with their present-day level of living.
- the more into the past the people go in their analysis, the more probable it is that they will revolt; people who do not know their history are less prone to revolutions than people who do.

References

1. Urenda, J., Kreinovich, V.: When revolutions happen: algebraic explanation. J. Uncertain Syst. **13**(2), 142–146 (2019)
2. Cialdini, R.B.: Influence: Science and Practice. Pearson, Boston (2009)
3. Davies, J.C.: Towards a theory of revolution. Am. Sociol. Rev. **27**, 5–19 (1962)
4. Davies, J.C.: The J-curve of rising and declining satisfactions as a cause of some great revolutions and a contained rebellion. In: Graham, H.D., Gurr, T.R. (eds.) Violence in America. Signet Books, New York (1969)
5. Fishburn, P.C.: Utility Theory for Decision Making. Wiley, New York (1969)
6. Kreinovich, V.: Decision making under interval uncertainty (and beyond). In: Guo, P., Pedrycz, W. (eds.) Human-Centric Decision-Making Models for Social Sciences, pp. 163–193. Springer (2014)
7. Luce, R.D., Raiffa, R.: Games and Decisions: Introduction and Critical Survey. Dover, New York (1989)
8. Nguyen, H.T., Kosheleva, O., Kreinovich, V.: Decision making beyond Arrow's 'impossibility theorem', with the analysis of effects of collusion and mutual attraction. Int. J. Intell. Syst. **24**(1), 27–47 (2009)
9. Raiffa, H.: Decision Analysis. McGraw-Hill, Columbus, Ohio (1997)
10. Nguyen, H.T., Kreinovich, V.: Applications of Continuous Mathematics to Computer Science. Kluwer, Dordrecht (1997)

Chapter 24
Application to Education: General

In this chapter and in the following chapter, we describe applications to education: a general application is described in this chapter, and a specific application is described in the following chapter.

In this chapter, we deal with the known fact that repetition enhances learning. The question is: when is a good time for this repetition? Several experiments have shown that immediate repetition of the topic leads to better performance on the resulting test than a repetition after some time. Recent experiments showed, however, that while immediate repetition leads to better results on the test, it leads to much worse performance in the long term, i.e., several years after the material have been studied. In this chapter, we use decision theory to provide a possible explanation for this unexpected phenomenon.

Results from this chapter first appeared in [1].

24.1   Problem: Is Immediate Repetition Good for Learning?

Repetitions are important for learning. A natural idea to make students better understand and better learn the material is to repeat this material—the more times we repeat, the better the learning results.

This repetition can be explicit—e.g., when we go over the material once again before the test. This repetition can be implicit—e.g., when we give the students a scheduled quiz on the topic, so that they repeat the material themselves when preparing for this quiz.

When should we repeat? The number of repetitions is limited by the available time. Once the number of repetitions is fixed, it is necessary to decide when should we have a repetition:

© The Author(s), under exclusive license to Springer Nature Switzerland AG 2022
J. C. Urenda and V. Kreinovich, *Algebraic Approach to Data Processing*,
Studies in Big Data 115, https://doi.org/10.1007/978-3-031-16780-5_24

- shall we have it immediately after the students have studied the material, or
- shall we have it after some time after this studying, i.e., after we have studied something else.

What was the recommendation until recently. Experiments have shown that repeating the material almost immediately after the corresponding topic was first studied—e.g., by giving a quiz on this topic—enhances the knowledge of this topic that the students have after the class as a whole. This enhancement was much larger than when a similar quiz—reinforcing the students' knowledge—was given after a certain period of time after studying the topic.

New data seems to promote the opposite recommendation. This idea has been successfully used by many instructors. However, a recent series of experiments has made many researchers doubting this widely spread strategy. Specifically, these experiments show that (see, e.g., [2] and references therein):

- while immediate repetition indeed enhances the amount of short-term (e.g., semester-wide) learning more than a later repetition,
- from the viewpoint of long-term learning—what the student will be able to recall in a few years (when he or she will start using this knowledge to solve real-life problems)—the result is opposite: delayed repetitions lead to much better long-term learning than the currently-fashionable immediate ones.

Why? The above empirical result is somewhat unexpected, so how can we explain it? We have partially explained the advantages of *interleaving*—a time interval between the study and the repetition—from the general geometric approach; see, e.g., [3, 4]. However, this explanation does not cover the difference between short-term and long-term memories.

So how can we explain this observed phenomenon? We can simply follow the newer recommendations, kind of arguing that human psychology is difficult, has many weird features, so we should trust whatever the specialists tell us. This may sound reasonable at first glance, but the fact that we have followed this path in the past and came up with what seems now to be wrong recommendation—this fact encourages us to take a pause, and first try to understand the observed phenomenon, and only follow it if it makes sense.

This is exactly the purpose of this chapter: to provide a reasonable explanation for the observed phenomenon.

24.2 Analysis of the Problem and the Resulting Explanation

Main idea: using decision theory. Our memory is limited in size. We cannot memorize everything that is happening to us. Thus, our brain needs to decide what to store in a short-term memory, what to store in a long-term memory, and what not to store at all.

How can we make this decision? As we have shown earlier, decision theory explains how we make decisions. Let us apply this to learning. If we learn the material, we spend some resources on storing it in memory. If we do not learn the material, we may lose some utility next tome when this material will be needed. So, whether we store the material in memory depends on for which of the two possible actions—to learn or not to learn—utility is larger (or equivalently, losses are smaller). Let us describe this idea in detail.

Notations. To formalize the above idea, let us introduce some notations.

- Let m denote the losses (= negative utility) needed to store a piece of material in the corresponding memory (short-term or long-term).
- Let L denote losses that occur when we need this material but do not have it in our memory.
- Finally, let p denote our estimate of the probability that this material will be needed in the corresponding time interval (short-term time interval for short-term memory or long-term time interval for long-term memory).

If we learn, we have loss m. If we do not learn, then the expected loss is equal to $p \cdot L$. We learn the material if the second loss of larger, i.e., if $p \cdot L > m$, i.e., equivalently, if $p > m/L$.

Comment. Sometimes, students underestimate the usefulness of the studied material, i.e., underestimate the value L. In this case, L is low, so the ratio m/L is high, and for most probability estimates p, learning does not make sense. This unfortunate situation can be easily repaired if we explain, to the students, how important this knowledge can be—and thus, make sure that they estimate the potential loss L correctly.

Discussion. For different pieces of the studied material, we have different ratios m/L. These ratios do not depend on the learning technique. As we will show later, the estimated probability p may differ for different learning techniques. So, if one technique consistently leads to higher values p, this means that, in general, that for more pieces of material we will have $p > m/L$ and thus, more pieces of material will be learned. So, to compare two different learning techniques, we need to compare the corresponding probability estimates p.

Let us formulate the problem of estimating the corresponding probability p in precise terms.

Towards a precise formulation of the probability estimation problem. In the absence of other information, to estimate the probability that this material will be needed in the future, the only information that our brain can use is that there were two moments of time at which we needed this material in the past:

- the moment t_1 when the material was first studied, and
- the moment t_2 when the material was repeated.

In the immediate repetition case, the moment t_2 was close to t_1, so the difference $t_2 - t_1$ was small. In the delayed repetition case, the difference $t_2 - t_1$ is larger.

Based on this information, the brain has to estimate the probability that there will be another moment of time during some future time interval. How can we do that?

Let us first consider a deterministic version of this problem. Before we start solving the actual probability-related problem, let us consider the following simplified deterministic version of this problem:

- we know the times $t_1 < t_2$ when the material was needed;
- we need to predict the next time t_3 when the material will be needed.

We can reformulate this problem in more general terms:

- we observed some event at moments t_1 and $t_2 > t_1$;
- based on this information, we want to predict the moment t_3 at which the same event will be observed again.

In other words, we need to have a function $t_3 = F(t_1, t_2) > t_2$ that produces the desired estimate.

What are the reasonable properties of this prediction function? The numerical value of the moment of time depends on what unit we use to measure time—e.g., hours, days, or months. It also depends on what starting point we choose for measuring time. We can measures it from Year 0 or—following Muslim or Buddhist calendars—from some other date.

If we replace the original measuring unit with the new one which is a times smaller, then all numerical values will multiply by a: $t \to t' = a \cdot t$. For example, if we replace seconds with milliseconds, all numerical values will multiply by 1000, so, e.g., 2 s will become 2000 ms. Similarly, if we replace the original starting point with the new one which is b units earlier, then the value b will be added to all numerical values: $t \to t' = t + b$. It is reasonable to require that the resulting prediction t_3 not depend on the choice of the unit and on the choice of the starting point. Thus, we arrive at the following definitions.

Definition 24.1 We say that a function $F(t_1, t_2)$ is *scale-invariant* if for every t_1, t_2, t_3, and $a > 0$, if $t_3 = F(t_1, t_2)$, then for $t_i' = a \cdot t_i$, we get $t_2' = F(t_1', t_2')$.

Definition 24.2 We say that a function $F(t_1, t_2)$ is *shift-invariant* if for every t_1, t_2, t_3, and b, if $t_3 = F(t_1, t_2)$, then for $t_i' = t_i + b$, we get $t_2' = F(t_1', t_2')$.

Proposition 24.1 *A function $F(t_1, t_2) > t_2$ is scale- and shift-invariant if and only if it has the form $F(t_1, t_2) = t_2 + \alpha \cdot (t_2 - t_1)$ for some $\alpha > 0$.*

Proof Let us denote $\alpha \overset{\text{def}}{=} F(-1, 0)$. Since $F(t_1, t_2) > t_2$, we have $\alpha > 0$. Let $t_1 < t_2$, then, due to scale-invariance with $a = t_2 - t_1 > 0$, the equality $F(-1, 0) = \alpha$ implies that $F(t_1 - t_2, 0) = \alpha \cdot (t_2 - t_1)$. Now, shift-invariance with $b = t_2$ implies that $F(t_1, t_2) = t_2 + \alpha \cdot (t_2 - t_1)$. The proposition is proven.

Discussion. Many physical processes are reversible: if we have a sequence of three events occurring at moments $t_1 < t_2 < t_3$, then we can also have a sequence of events at times $-t_3 < -t_2 < -t_1$. It is therefore reasonable to require that:

- if our prediction works for the first sequence, i.e., if, based on t_1 and t_2, we predict t_3,
- then our prediction should work for the second sequence as well, i.e. based on $-t_3$ and $-t_2$, we should predict the moment $-t_1$.

Let us describe this requirement in precise terms.

Definition 24.3 We say that a function $F(t_1, t_2)$ is *reversible* if for every t_1, t_2. and t_3, the equality $F(t_1, t_2) = t_3$ implies that $F(-t_3, -t_2) = -t_1$.

Proposition 24.2 *The only scale- and shift-invariant reversible function $F(t_1, t_2)$ is the function $F(t_1, t_2) = t_2 + (t_2 - t_1)$.*

Comment. In other words, if we encounter two events separated by the time interval $t_2 - t_1$, then the natural prediction is that the next such event will happen after exactly the same time interval.

Proof In view of Proposition 24.1, all we need to do is to show that for a reversible function we have $\alpha = 1$. Indeed, for $t_1 = -1$ and $t_2 = 0$, we get $t_3 = \alpha$. Then, due Proposition 24.1.1, we have $F(-t_3, -t_2) = F(-\alpha, 0) = 0 + \alpha \cdot (0 - (-\alpha)) = \alpha^2$. The requirement that this value should be equal to $-t_1 = 1$ implies that $\alpha^2 = 1$, i.e., due to the fact that $\alpha > 0$, that $\alpha = 1$. The proposition is proven.

From simplified deterministic case to the desired probabilistic case. In practice, we cannot predict the actual time t_3 of the next occurrence, we can only predict the *probability* of different times t_3. Usually, the corresponding uncertainty is caused by a joint effect of many different independent factors. It is known that in such situations, the resulting probability distribution is close to Gaussian—this is the essence of the Central Limit Theorem which explains the ubiquity of Gaussian distributions; see, e.g., [5]. It is therefore reasonable to conclude that the distribution for t_3 is Gaussian, with some mean μ and standard deviation σ.

There is a minor problem with this conclusion; namely:

- Gaussian distribution has non-zero probability density for all possible real values, while
- we want to have only values $t_3 > t_2$.

This can be taken into account if we recall that in practice, values outside a certain $k\sigma$-interval $[\mu - k \cdot \sigma, \mu + k \cdot \sigma]$ have so little probability that they are considered to be impossible. Depending on how low we want this probability to be, we can take $k = 3$, or $k = 6$, or some other value k. So, it is reasonable to assume that the lower endpoint of this interval corresponds to t_2, i.e., that $\mu - k \cdot \sigma = t_2$. Hence, for given

t_1 and t_2, once we know μ, we can determine σ. Thus, to find the corresponding distribution, it is sufficient to find the corresponding value μ.

As this mean value μ, it is reasonable to take the result of the deterministic prediction, i.e., $\mu = t_2 + (t_2 - t_1)$. In this case, from the above formula relating μ and σ, we conclude that $\sigma = (t_2 - t_1)/k$.

Finally, an explanation. Now we are ready to explain the observed phenomenon.

In the case of immediate repetition, when the difference $t_2 - t_1$ is small, most of the probability—close to 1—is located is the small vicinity of t_1, namely in the $k\sigma$ interval which now takes the form $[t_2, t_2 + 2(t_2 - t_1)]$. Thus, in this case, we have:

- (almost highest possible) probability $p \approx 1$ that the next occurrence will have in the short-term time interval and
- close to 0 probability that it will happen in the long-term time interval.

Not surprisingly, in this case, we get:

- a better short-term learning than for other learning strategies, but
- we get much worse long-term learning.

In contrast, in the case of delayed repetition, when the difference $t_2 - t_1$ is large, the interval $[t_2, t + 2(t_2 - t_1)]$ of possible values t_3 spreads over long-term times as well. Thus, here:

- the probability p to be in the short-time interval is smaller than the value ≈ 1 corresponding to immediate repetition, but
- the probability to be in the long-term interval is larger that the value ≈ 0 corresponding to immediate repetition.

As a result, for this learning strategy:

- we get worse short-term learning but
- we get much better long-term learning,

exactly as empirically observed.

References

1. Bokati, L., Urenda, J., Kosheleva, O., Kreinovich, V.: Why immediate repetition is good for short-term learning results but bad for long-term learning: explanation based on decision theory. In: Ceberio, M., Kreinovich, V. (eds.) How Uncertainty-Related Ideas Can Provide Theoretical Explanation for Empirical Dependencies, pp. 27–35. Springer, Cham, Switzerland (2021)
2. Epstein, D.: Range: Why Generalists Triumph in the Specialized World. Riverhead Books, New York (2019)
3. Kosheleva, O., Villaverde, K.: How Interval and Fuzzy Techniques Can Improve Teaching. Springer, Cham, Switzerland (2018)
4. Lerma, O., Kosheleva, O., Kreinovich, V.: Interleaving enhances learning: a possible geometric explanation. Geombinatorics **24**(3), 135–139 (2015)
5. Sheskin, D.J.: Handbook of Parametric and Non-parametric Statistical Procedures. Chapman & Hall/CRC, London, UK (2011)

Chapter 25
Application to Education: Specific

To many students, the notion of a derivative seems unrelated to any previous mathematics—and is, thus, difficult to study and to understand. In this chapter, we show that this notion can be naturally derived from a more intuitive notion of invariance.

Results from this chapter first appeared in [1].

25.1 Problem: Why Derivative

To a student studying mathematics, the notion of the derivative seems to appear out of nowhere, without any explanation and without any reasonable relation with previously studied mathematical notions. This un-relatedness may be one of the reasons why calculus is so difficult for many students, even for those who have successfully studied previous mathematical subjects.

In this chapter, we show that the notion of the derivative can be explained by the natural ideas of shift- and scale-invariance. We hope that this explanation will make this notion more natural—and thus, easier to learn.

25.2 Invariance Naturally Leads to the Derivative

Let us start the construction. Now, we are ready to show that the natural notions of invariance indeed lead to the expression for the derivative. We will do it step-by-step, adding more invariance requirements as we go.

We have a function $y = f(x)$. Based on the values of this function, we want to build a new auxiliary function $g(x)$. Let us consider the simplest case when at each point x, the value of the new function $g(x)$ will depend only on two values of the original function $f(x)$. In other words, we consider the case when

© The Author(s), under exclusive license to Springer Nature Switzerland AG 2022   129
J. C. Urenda and V. Kreinovich, *Algebraic Approach to Data Processing*,
Studies in Big Data 115, https://doi.org/10.1007/978-3-031-16780-5_25

$$g(x) = F(f(p_1(x)), f(p_2(x)),\qquad(25.2)$$

where:

- $p_1(x)$ and $p_2(x)$ describe how these two points depend on x, and
- $F(y, z)$ is an algorithm that transforms the corresponding two values of the function $f(x)$ into the value $g(x)$ of the new function.

First invariance requirement: invariance with respect to x-shifts. The first natural invariance requirement that we will impose is x-*shift-invariance*: if we use a different starting point for measuring x, the expressions for the corresponding dependencies $p_1(x)$ and $p_2(x)$ should not change. Let us describe this requirement in precise terms.

Each expression $p_i(x)$ describes how the value of the corresponding point x_i in the original x-scale depends on the value of the parameter x in the same scale. If we change the starting point, then each original value x will take the new form $\widetilde{x} = x + b$, so that $x = \widetilde{x} - b$, and the point $x_i = p_i(x)$ at which we should compute $f(x)$ will take a new form $\widetilde{x}_i = p_i(x) + b$. Substituting the expression $x = \widetilde{x} - b$ into this formula, we conclude that in the new scale, the dependence of the corresponding point $\widetilde{x}_i$ on $\widetilde{x}$ should take the form $\widetilde{x}_i = p_i(\widetilde{x} - b) + b$. Invariance means that this dependence should be expressed by the same formula as in the original scale, i.e., we should have $\widetilde{x}_i = p_i(\widetilde{x})$.

Comparing these two expressions, we conclude that $p_i(\widetilde{x} - b) + b = p_i(\widetilde{x})$ for all b and $\widetilde{x}_i$. In particular, for $b = \widetilde{x}$, we conclude that $p_i(\widetilde{x}) = \widetilde{x} + p_i(0)$. Thus, due to this invariance requirements, each function $p_i(x)$ has the form $p_i(x) = x + \text{const}$. Let us denote the corresponding constant by c_i. Then, we have $p_i(x) = x + c_i$, and the formula (25.2) takes the form

$$g(x) = F(f(x + c_1), f(x + c_2)).\qquad(25.3)$$

This expression is simpler than the original expression (25.2):

- in the original expression (25.2), we had three unknown functions $F(y_1, y_2)$, $p_1(x)$, and $p_2(x)$, while
- now, we have only one unknown function $F(y_1, y_2)$ and two unknown numbers c_1 and c_2.

Second invariance requirement: invariance with respect to y-shifts. Another reasonable requirement is that the values $g(x)$ should not change if we simply change the starting point for measuring y. As we have mentioned earlier, this change simply adds the same constant b to all the y-values—i.e., in our case, to both values of the function $f(x)$. Thus, instead of the original value $F(f(x + c_1), f(x + c_2))$, we will have a new value $F(f(x + c_1) + b, f(x + c_2) + b)$. Invariance means that these two values must coincide, i.e., that we should have

$$g(x) = F(f(x + c_1), f(x + c_2)) = F(f(x + c_1) + b, f(x + c_2) + b)$$

for all x and b. In particular, for $b = -f(x + c_2)$, we have

$$g(x) = F(f(x + c_1), f(x + c_2)) = F(f(x + c_1) - f(x + c_2), 0),$$

i.e., equivalently, that

$$g(x) = G(f(x + c_1) - f(x + c_2)), \tag{25.4}$$

where we denoted $G(y) \overset{\text{def}}{=} F(y, 0)$. This expression is even simpler than the expression (25.3):

- in the expression (25.3), we had an unknown function $F(y_1, y_2)$ of two variables, while
- now, we have only an unknown function $G(y)$ of one variable.

Next invariance requirement: invariance with respect to y-scaling. Another natural invariance requirement is that the dependence (25.4) should not change if we change the unit in which we measure y-values like $f(x)$ or $g(x)$. In other words, if we have the expression (25.4) and we replace $f(x + c_i)$ with $\tilde{f}(x + c_i) = a \cdot f(x + c_i)$ and $g(x)$ with $\tilde{g}(x) = c \cdot g(x)$, then we should have the same relation between the re-scaled values, i.e., we should have

$$\tilde{g}(x) = G\left(\tilde{f}(x + c_1) - \tilde{f}(x + c_2)\right).$$

In other words, we should have

$$G(\lambda \cdot f(x + c_1) - \lambda \cdot f(x + c_2)) = \lambda \cdot g(x) = \lambda \cdot G(f(x + c_1) - f(x + c_2)),$$

i.e.,

$$G(\lambda \cdot (f(x + c_1) - f(x + c_2))) = \lambda \cdot G(f(x + c_1) - f(x + c_2)).$$

Since the difference $z \overset{\text{def}}{=} f(x + c_1) - f(x + c_2)$ can take any possible real value, we thus have

$$G(\lambda \cdot z) = \lambda \cdot G(z).$$

In particular, for $z = 1$, we conclude that $G(\lambda) = \lambda \cdot G(1)$, i.e., that $G(\lambda) = K \cdot \lambda$, where we denoted $K \overset{\text{def}}{=} G(1)$. For this function $G(z)$, the formula (25.4) takes an even simpler form

$$g(x) = K \cdot (f(x + c_1) - f(x + c_2)). \tag{25.5}$$

We can have different expressions like that, for different values c_1 and c_2. In general, the coefficient K may depend on which values c_1 and c_2 we select, so we get

$$g(x) = K(c_1, c_2) \cdot (f(x + c_1) - f(x + c_2)). \tag{25.5a}$$

Which values $K(c_1, c_2)$ should we choose? In general, the value of the expression (25.5a) changes when we change the values c_1 and c_2. In particular, this is true even if we consider the invariant dependencies $f(x)$—which, as we have shown in the previous section, correspond to linear functions $f(x) = a \cdot x + b$.

For a linear function $f(x) = a \cdot x + b$, the expression (25.5a) takes the form

$$g(x) = K(c_1, c_2) \cdot ((a \cdot (x + c_1) + b) - (a \cdot (x + c_2) + b)) = K(c_1, c_2) \cdot a \cdot (c_1 - c_2) = a \cdot c,$$

where we denoted $c \stackrel{\text{def}}{=} K(c_1, c_2) \cdot (c_1 - c_2)$.

Thus, it is possible to select the coefficient $K(c_1, c_2)$ in such a way that for linear functions, the resulting value $g(x)$ will not depend on the selection of c_1 and c_2. Namely, to make sure the product $c = K(c_1, c_2) \cdot (c_1 - c_2)$ remains the same for all c_1 and c_2, we should select $K(c_1, c_2) = \dfrac{c}{c_1 - c_2}$. In this case, the expression (25.5a) takes the form

$$g(x) = c \cdot \frac{f(x + c_1) - f(x + c_2)}{c_1 - c_2}. \tag{25.6}$$

Which values c_1 and c_2 should we choose? A reasonable idea is to consider *local* characteristics, i.e., characteristics $g(x)$ that depend only the values of the original function $f(x)$ in a small vicinity of the point x: e.g., in the ε-vicinity of all the points which are ε-close to x. Thus, we consider cases when the values c_1 and c_2 are small: e.g., $|c_i| \leq \varepsilon$ for $i = 1, 2$.

As we consider the smaller and smaller neighborhoods, the values c_i tend to 0 and thus, we get the value

$$g(x) = c \cdot \lim_{c_1, c_2 \to 0} \frac{f(x + c_1) - f(x + c_2)}{c_1 - c_2}. \tag{25.7}$$

Modulo a multiplicative constant c, this is exactly the derivative—i.e., exactly the expression that we wanted to explain.

How can we describe the above expression for the derivative in a more standard form. While the expression (25.7) is equal to the derivative (modulo c), it is *different* from the standard definitions of the derivative. We can make it closer if we impose an additional invariance requirement: that the formula (25.6) (and thus, the formula (25.7) should not change if we replace c with $\widetilde{x} = -x$. In this case, we have $x = -\widetilde{x}$, so, instead of the original function $f(x)$, we get a new function $\widetilde{f}(\widetilde{x}) \stackrel{\text{def}}{=} f(-\widetilde{x})$. If we apply the formula (25.6) to this new function, we get the expression

$$\widetilde{g}(\widetilde{x}) = c \cdot \frac{\widetilde{f}(\widetilde{x} + c_1) - \widetilde{f}(\widetilde{x} + c_2)}{c_1 - c_2}.$$

Invariance means that, when we substitute the formulas for $\widetilde{f}(z)$ and for $\widetilde{x} = -x$ into this expression, we should get the same formula (25.6). Here,

$$\widetilde{f}(\widetilde{x} + c_i) = f(-(\widetilde{x} + c_i)) = f(-(-x + c_i)) = f(x - c_i),$$

thus the desired equality takes the form:

$$\frac{f(x - c_1) - f(x - c_2)}{c_1 - c_2} = \frac{f(x + c_1) - f(x + c_2)}{c_1 - c_2}.$$

This equality should be satisfied for all possible functions $f(x)$. Thus, the left-hand side should use the values of the function $f(x)$ at exactly the same two points as the right-hand side. The only two possible options for this equality are:

- the case when $c_1 = -c_1$ and $c_2 = -c_2$, and
- the case when $c_1 = -c_2$ and $c_2 = -c_1$.

In the first case, we get $c_1 = c_2 = 0$ and thus, $g(x)$ is always equal to 0. The only non-trivial case is the second case, in which case (25.6) takes the form

$$g(x) = c \cdot \frac{f(x + h) - f(x - h)}{2h}, \qquad (25.8)$$

where we denoted $h \overset{\text{def}}{=} c_1$. In this case, the limit expression (25.7) turns into one of the often-used versions of the standard definition of the derivative:

$$g(x) = c \cdot \lim_{h \to 0} \frac{f(x + h) - f(x - h)}{2h}. \qquad (25.9)$$

Reference

1. Urenda, J., Kosheleva, O., Kreinovich, V.: Why derivative: invariance-based explanation. Math Struct Modeling **52**, 134–140 (2019)

Chapter 26
Application to Mathematics: Why Necessary Conditions Are Often Sufficient

In this chapter, we provide an application to mathematics. In many graph-related problems, an obvious necessary condition is often also sufficient. This phenomenon is so ubiquitous that it was even named TONCAS, after the first letters of the phrase describing this phenomenon. In this chapter, we provide a possible explanation for this phenomenon.

The results of this chapter first appeared in [1].

26.1 Formulation of the Problem

TONCAS phenomenon: an example. When is a graph planar? In precise terms, when can a graph be embedded in a plane, i.e., represented by a graph in which edges intersect only at vertices? Clearly, if a graph contains a subgraph K_5 is which all five vertices are connected to each other, it cannot be embedded into a plane. Similarly, a planar graph cannot contain a subgraph $K_{3,3}$ that has two groups of 3 vertices, so that each of each vertex from the first group is connected with each vertex from the second group. Interesting, these two necessary conditions are sufficient: if a graph does not contain any subgraphs isomorphic to K_5 or to $K_{3,3}$, then this graph is planar; see, e.g., [2].

Another known example is checking whether a given lattice is distributive: this is equivalent to requiring that no sublattice is isomorphic to one of the prohibitive 5- and 6-vertices sublattices; see, e.g., [3].

TONCAS: general phenomenon. It turns out that this phenomenon is ubiquitous in graph theory and in related areas: in many such cases, the obvious necessary condition is also sufficient. This is known as the TONCAS phenomenon, after the first letters of the words that describe this phenomenon.

Why? A natural question is: why? How can we explain that for many natural properties, we have this phenomenon?

J. C. Urenda and V. Kreinovich, *Algebraic Approach to Data Processing*, Studies in Big Data 115, https://doi.org/10.1007/978-3-031-16780-5_26

26.2   Analysis of the Problem

Let us formulate the TONCAS phenomenon in general terms. For simplicity, let us consider properties like embedding-related ones, in which, if the entire graph has this property, then all its subgraphs also satisfy the same property. In particular, if a graph satisfies the given property, then all its subgraphs with n or fewer vertices has this property. The TONCAS phenomenon can be described as follows: there is a reasonably small value n_0 such that if the desired property is satisfied for all subgraphs with n_0 or fewer vertices, then this propety is satisfied by the graph itself.

For planarity, as we have mentioned, $n_0 = 6$: if all subgraphs with 6 or fewer vertices are planar, this means that none of them is isomorphic to K_5 or to $K_{3,3}$ and thus, the whole graph is indeed planar. The similar bound $n_0 = 6$ holds for checking whether a lattice is distributive.

Describing TONCAS phenomenon in precise terms. Let us denote that by $P(n)$ the condition that all subgraphs with n or fewer vertices satisfy the desired property. Then, clearly, $P(n+1)$ implies $P(n)$.

In these terms, the fact that the whole graph—no matter how many vertices it has—satisfies the desired property means that the condition $P(n)$ holds for all n.

Thus, the TONCAS phenomenon means that there exists a value n_0 such that for each reasonable (= not abnormal) predicate P for which $P(n+1)$ implies $P(n)$, the condition $P(n_0)$ implies $\forall n \, P(n)$.

Caution. Of course, for each n_0, we can always find some artificial predicates $P(n)$ for which, for some graphs, we have $P(n_0)$ but $P(n)$ is not true for some $n > n_0$. This will happen, e.g., if the original desired graph property is that the graph has $\leq n_0$ vertices. For this property, clearly $P(n_0)$ is true, but also clearly, $P(n_0 + 1)$ is not true for any graph with more than n_0 vertices.

So, to explain the TONCAS property, we cannot just ignore the words "reasonable" and "not abnormal", we need to formalize them.

26.3   How Can We Formalize What Is Not Abnormal

Let us use the experience of statistical physics. Many real-life phenomena are probabilistic. From the purely mathematical viewpoint, if we have, e.g., a Gaussian (normal) distribution on a real line, with 0 mean and standard deviation 1, then, since the probability density of the normal distribution is always positive, for every n—no matter how large it is—there is a positive probability that the random value will be larger than n. However, this is *not* how physicists reason; see, e.g., [4, 5].

For example, from the purely mathematical viewpoint, it is possible that, due to Brownian motion, a kettle placed on a cold stove will start boiling by itself—or that randomly moving molecules in a human body start moving in the same direction and the person will float into the air. A mathematician may say that this will happen if we wait a sufficiently long time—very long time, since the probabilities of these events are extremely small. However, this is *not* what a physicist will say. A physicist will simply claim that these events are *not* possible. In general, a physicist will say that if an event has a very low probability, then this even simply cannot happen.

How can we describe this physicists' reasoning in precise terms. Of course, we *cannot* simply fix a small number p_0 and claim that any event with probability $\leq p_0$ is not possible. Indeed, in this case, for sufficiently large N—for which $2^{-N} \leq p_0$— we would come up with an awkward conclusion that it is impossible to flip a coin N times, since each of the 2^N possible sequences of heads and tails has the same probability $2^{-N} \leq p_0$.

What we *can* conclude is that if we have a definable sequence of events $E_1 \supseteq E_2 \supseteq \cdots$ for which the probabilities $p(E_n)$ tend to 0, then at some point, the corresponding probability will be so small than the corresponding event E_n will simply not be possible. In other words, the class $\mathscr{R}$ of all *reasonable* (*not abnormal*) observations should satisfy the following property: if for some definable sequence $E_n \supseteq E_{n+1}$, we have $p(E_n) \rightarrow 0$, then there exists a value n_0 for which $\mathscr{R} \cap E_{n_0} = \emptyset$; see, e.g., [6–8].

What if we do not know probabilities? In statistical physics, we know the probabilities, but in graph-related situations, there is no natural way to assign probabilities. However, we can still use the above description if we take into account that there is a natural case when we can guarantee $p(E_n) \rightarrow 0$ for all possible probability distributions: namely, the case when the intersection $\bigcap_n E_n$ of all the sets E_n is empty.

In this case, we arrive at the following description: the class $\mathscr{R}$ of all *reasonable* (*not abnormal*) observations should satisfy the following property: if for some definable sequence $E_n \supseteq E_{n+1}$, we have $\bigcap_n E_n = \emptyset$, then there exists a value n_0 for which $\mathscr{R} \cap E_{n_0} = \emptyset$; see, e.g., [6–8].

26.4 Resulting Explanation of the TONCAS Phenomenon

Now, we are ready to explain the TONCAS phenomenon. We consider predicates $P(n)$ whose parameter n is a natural number. Let us call a predicate *monotonic* if for each n, $P(n + 1)$ implies $P(n)$. We will prove that there exists a natural number n_0 such that if the monotonic predicate P is *not abnormal*, then $P(n_0)$ implies $\forall n \, P(n)$.

Indeed, let us consider the sets

$$E_n \overset{\text{def}}{=} \{P : P(1) \& \ldots \& P(n) \& \exists m \, \neg P(m)\}.$$

Clearly, here $E_n \supseteq E_{n+1}$ and $\bigcap_n E_n = \emptyset$. Thus, by the above definition of not-abnormality, there exists an n_0 for which none of the not-abnormal monotonic predicates is contained in the set E_{n_0}. By the definition of the set E_{n_0}, this means that if we have $P(n_0)$ (and thus, due to monotonicity, $P(1)$ & ... & $P(n_0)$), then we cannot have $\exists m \, \neg P(m)$ and thus, we have $\forall m \, P(m)$.

This is exactly the TONCAS phenomenon—which is, therefore, justified.

References

1. Urenda, J.C., Kreinovich, V.: Why the obvious necessary condition is (often) also sufficient (TONCAS): an explanation of the phenomenon. Math. Struct. Model. **52**, 93–96 (2019)
2. West, D.B.: Introduction to Graph Theory. Pearson, New York (2000)
3. Birkhoff, G.: Lattice Theory. American Mathematical Society, Providence (1967)
4. Feynman, R., Leighton, R., Sands, M.: The Feynman Lectures on Physics. Addison Wesley, Boston (2005)
5. Thorne, K.S., Blandford, R.D.: Modern Classical Physics: Optics, Fluids, Plasmas, Elasticity, Relativity, and Statistical Physics. Princeton University Press, Princeton (2017)
6. Finkelstein, A.M., Kreinovich, V.: Impossibility of hardly possible events: physical consequences. In: Abstracts of the 8th International Congress on Logic. Methodology, and Philosophy of Science, Moscow, vol. 5, no. 2, pp. 23–25 (1987)
7. Kreinovich, V.: Toward formalizing non-monotonic reasoning in physics: the use of Kolmogorov complexity. Revista Iberoamericana de Inteligencia Artificial **41**, 4–20 (2009)
8. Kreinovich, V., Finkelstein, A.M.: Towards applying computational complexity to foundations of physics. Notes of Mathematical Seminars of St. Petersburg Department of Steklov Institute of Mathematics, vol. 316, pp. 63–110 (2004); Reprinted in J. Math. Sci. **134**(5), 2358–2382 (2006)

Chapter 27
Data Processing: Neural Techniques, Part 1

In addition to specific *applications*, we also apply algebraic techniques to *computational methods* leading to such applications. At present, the most promising data processing techniques are techniques corresponding to machine learning, especially to neural networks, in particular, to deep neural networks. These techniques are very successful, but they are not perfect.

One of the problems is that these techniques are largely heuristic, many of their features lack a solid theoretical foundation—a foundation that would increase our trust in these techniques. In this chapter, we use algebraic approach to provide a justification for some of these features—and we show that this justification also explains a successful heuristic modification of some of these features.

Results from this chapter first appeared in [1].

27.1 Machine Learning Is Needed to Analyze Complex Systems

For some simple systems, we know the equations that describe the system's dynamics. These equations may be approximate, but they are often good enough.

With more complex systems (such as systems of systems), this is often no longer the case. Even when we have a good approximate model for each subsystem, the corresponding inaccuracies add up, and the resulting model of the whole system is too inaccurate to be useful. For real-life systems like a city or a big plant, it is therefore often not possible to predict the system's behavior based only on approximate models of subsystems. In addition to these models, we also need to use the records of the actual system's behavior when making such predictions.

Techniques that use the previous behavior of a system to predict its future behavior are known as *machine learning* techniques.

© The Author(s), under exclusive license to Springer Nature Switzerland AG 2022
J. C. Urenda and V. Kreinovich, *Algebraic Approach to Data Processing*,
Studies in Big Data 115, https://doi.org/10.1007/978-3-031-16780-5_27

27.2 Neural Networks and Deep Learning: A Brief Reminder

At present, the most empirically successful machine learning techniques are neural networks—techniques that simulate how data in processed in our brains. In this technique, signal go through several sequential stages of data processing—known as *layers*. Each layer consists of several *neurons*, each of which transforms the inputs signals $x_1, \ldots, x_n$ to this layer this into a new signal $y = s\left(\sum_{i=1}^{n} w_i \cdot x_i + w_0\right)$. Here, the coefficient w_i (called *weights*) are to be determined during training, and $s(z)$ is a non-linear function called *activation function*.

Traditional neural networks use only two layers, the first of which uses the so-called *sigmoid* or *logistic activation function*

$$s(z) = \frac{1}{1 + \exp(-z)}, \tag{27.1}$$

and the second one contains only linear transformations.

More empirically successful are *deep* neural networks that have more layers. In deep neural networks, some layers use sigmoid activation functions, but most of the layers use a different activation function $s(z) = \max(0, z)$ known as *rectified linear* activation function.

Deep learning. At present, the most efficient machine learning technique is *deep learning*, i.e., the use of multi-layer neural networks; see, e.g., [2]. In general, on layer of a neural network, we transform signals $x_1, \ldots, x_n$ into a new signal $y = s\left(\sum_{i=1}^{n} w_i \cdot x_i + w_0\right)$, where the coefficient w_i (called *weights*) are to be determined during training, and $s(z)$ is a non-linear function called *activation function*.

Activation functions used in deep learning. Most multi-layer neural networks used in deep learning utilize rectified linear neurons, i.e., neurons that use the activation function $s(z) = \max(z, 0)$ known as *rectified linear* function.

Algebraic approach shows (see, e.g., [3, 4]) that if we want to use the exact same activation function for all the neurons, then the rectified linear function is indeed a reasonable choice.

However, empirical results show that in some applications, it is more advantageous to use different activation functions for different neurons—i.e., select a family of activation functions instead, and select the parameters of activation functions of different neurons during training. Specifically, this was shown for a special family of *squashing* activation functions that contain rectified linear neurons as a particular case; see, e.g., [5–7]. Functions from this family have the form

$$S_{a,\lambda}^{(\beta)}(z) = \frac{1}{\lambda \cdot \beta} \cdot \ln \frac{1 + \exp(\beta \cdot z - (a - \lambda/2))}{1 + \exp(\beta \cdot z - (a + \lambda/2))}. \tag{27.2}$$

27.3 Why Traditional Neural Networks

Let us first recall why traditional neural networks appeared in the first place; see, e.g., [4].

The main reason, in our opinion, was that computers were too slow. A natural way to speed up computations is to make several processors work in parallel—so that each processor only handles a simple task, not requiring too much computation time.

For processing data, the simplest possible functions to compute are linear functions. However, we cannot only use linear functions—because then, no matter how many linear transformations we apply one after another, we will only get linear functions, and many real-life dependencies are nonlinear. So, we need to supplement linear computations with some nonlinear ones. In general, the fewer inputs, the faster the computations. Thus, the fastest to compute are functions with one input, i.e., functions of one variable. So, we end up with a parallel computational device that has linear processing units (L) and nonlinear processing units (NL) that compute functions of one variable. First, the input signals come to a layer of such devices, then the results of this layer go to another layer, etc. The fewer layers we have, the faster the computations.

It can be shown (see, e.g., [4]) that 1-layer schemes (L or NL) and 2-layer schemes (L-NL, linear layer followed by non-linear layer, or NL-L) are not sufficient to approximate any possible dependence. Thus, we need at least 3-layer networks— and 3-layer networks can be proven to be sufficient. In a 3-layer network, we cannot have two linear layers or two nonlinear layers following each other—that would be equivalent to having one layer since, e.g., a composition of two linear functions is also linear. Thus, we have only two options: L-NL-L and NL-L-NL. Since linear transformations are faster to compute, the fastest scheme is L-NL-L. In this scheme:

- first, each neuron k in the L layer combines the inputs into a linear combination
$$z_k = \sum_{i=1}^{n} w_{ki} \cdot x_i + w_{k0};$$
- then, in the next layer, each such signal is transformed into $y_k = s_k(z_k)$ for some non-linear function; and
- finally, in the last linear layer, we form a linear combination of the values y_k:
$$y = \sum_{k=1}^{K} W_k \cdot y_k + W_0.$$

The resulting transformation takes the form

$$y = \sum_{k=1}^{K} W_k \cdot s_k \left(\sum_{i=1}^{n} w_{ki} \cdot x_i + w_{k0} \right) + W_0.$$

Usually, we use the same function $s(z)$ for all transformations, so we get

$$y = \sum_{k=1}^{K} W_k \cdot s \left(\sum_{i=1}^{n} w_{ki} \cdot x_i + w_{k0} \right) + W_0.$$

This is indeed the usual formula of the traditional neural network.

27.4   Why Sigmoid Activation Function: Idea

In real life, signals come with noise, in particular, with background noise that, in effect, adds a constant to all the measured signals. We can try to get rid of this noise by subtracting the corresponding constant, i.e., by replacing the original numerical values x_i with a corrected value $x_i - n_i$. After this correction, instead of the original value z_k, we get a corrected value

$$z'_k = \sum_{i=1}^{n} w_{ki} \cdot (x_i - n_i) + w_{k0} = z_k - h'_k,$$

where we denoted $h'_k \stackrel{\text{def}}{=} \sum_{i=1}^{n} w_{ki} \cdot n_i$.

The trouble is that we do not know the exact value of this constant—otherwise, this noise would not be a problem. So, depending on our estimate, we may subtract different values n_i and thus, different values h'_k. If we change from one value h'_k to another one h''_k, then the resulting value of z_k is shifted by the difference $h_k \stackrel{\text{def}}{=} h'_k - h''_k$, namely, $z''_k = z'_k + h_k$, exactly the same formula as for the shift corresponding to the change in the starting point.

Since we do not know what shift is the best, all shifts within a certain range are equally possible. It is therefore reasonable to require that the formula $y = s(z)$ for the nonlinear activation function should work for all possible shifts. In other words, as we mentioned in the previous section, if we shift from z to $z' = z + h$, then we should satisfy the exact same formula $y' = s(z')$—probably for an appropriately transformed value y.

In the previous section, we also mentioned that all possible transformations should be fractional-linear. Thus, for each possible shift h, the value $s(z') = s(z + h)$ should be obtained from $s(z)$ by an appropriate fractional-linear transformation:

$$s(z + h) = \frac{a(h) \cdot s(z) + b(h)}{c(h) \cdot s(z) + d(h)}. \tag{27.3}$$

Let us show that this implies the sigmoid.

27.5 Why Sigmoid—Derivation

For $h = 0$, we should have $s(z + h) = s(z)$, thus, we should have $d(0) \neq 0$. It is reasonable to require that the function $d(h)$ is continuous. In this case, $d(h)$ is different from 0 for all small h. Then, we can divide both numerator and denominator of the formula (27.3) by $d(h)$ and get a simpler formula, with only three functions of h:

$$s(z + h) = \frac{A(h) \cdot s(z) + B(h)}{C(h) \cdot s(z) + 1}, \qquad (27.4)$$

where we denoted $A(h) = a(h)/d(h)$, $B(h) = b(h)/d(h)$, and $C(h) = c(h)/d(h)$. For $h = 0$, we have $s(z + h) = s(z)$, so $A(h) = 1$ and $B(h) = C(h) = 0$.

It is also reasonable to require that the activation function $s(z)$ be smooth. We also want it to be defined for all z.

Smoothness requirement comes from the fact that on each interval, every continuous function can be approximated, with any desired accuracy, by a smooth one—even by a polynomial. So we can always get non-smooth functions as limits of smooth ones.

Multiplying both sides of the formula (27.4) by the denominator and moving the term $s(z + h) \cdot C(h)$ to the right-hand side, we get the following formula:

$$s(z + h) = A(h) \cdot s(z) + B(h) - s(z + h) \cdot C(h).$$

In particular, if we take three different values $z = z_1$, $z = z_2$, and $z = z_3$, then, for each h, we get the following system of three linear equations for determining the three values $A(h)$, $B(h)$, and $C(h)$:

$$s(z_1 + h) = A(h) \cdot s(z) + B(h) - s(z_1 + h) \cdot C(h);$$

$$s(z_2 + h) = A(h) \cdot s(z) + B(h) - s(z_2 + h) \cdot C(h);$$

$$s(z_3 + h) = A(h) \cdot s(z) + B(h) - s(z_3 + h) \cdot C(h).$$

Due to Cramer's rule, the solution to this system is a ratio of two determinants, i.e., of two polynomials of the coefficients and is, thus, a smooth function of the values $s(z_i + h)$. Since the function $s(z)$ is smooth, we conclude that all three functions $A(h)$, $B(h)$, and $C(h)$ are also smooth. Thus, we can differentiate both sides of the Eq. (27.4) by h and get

$$s'(z + h) = \frac{N(h)}{(C(h) \cdot s(z) + 1)^2},$$

where

$$N(h) \stackrel{\text{def}}{=} (A'(h) \cdot s(z) + B'(h)) \cdot (C(h) \cdot s(z) + 1) -$$

$$(A(h) \cdot s(z) + B(h)) \cdot (C'(h) \cdot s(z)).$$

In particular, for $h = 0$, taking into account that $A(h) = 1$ and $B(h) = C(h) = 0$, we conclude that

$$s'(z) = a_0 + a_1 \cdot s(z) + a_2 \cdot (s(z))^2, \tag{27.5}$$

where we denoted $a_0 = B'(0)$, $a_1 = A'(0)$, and $a_2 = -C'(0)$, i.e., $\dfrac{ds}{dz} = a_0 + a_1 \cdot s + a_2 \cdot s^2$. If we move all the terms related to s to the left-and side and all the terms related to z to the right-hand side, we get the following formula:

$$\frac{ds}{a_0 + a_1 \cdot s + a_2 \cdot s^2} = dz. \tag{27.6}$$

Let us show how we can integrate both sides of this formula and get an explicit expression of $z(s)$, and how based on this expression, we can find the explicit formula for the dependence of s on z.

The generic case is when $a_2 \neq 0$. In this case, we can multiply both sides of the formula (27.6) by a_2 and get

$$\frac{ds}{\dfrac{a_0}{a_2} + \dfrac{a_1}{a_2} \cdot s + s^2} = a_2 \cdot dz. \tag{27.7}$$

The quadratic form in the denominator of the left-hand side can be represented as $(s + p)^2 + q$, where $p = \dfrac{a_1}{2a_2}$ and $q = \dfrac{a_0}{a_2} - p^2$. Thus, the formula (27.7) takes the form

$$\frac{ds}{(s + p)^2 + q} = a_2 \cdot dz. \tag{27.8}$$

So, for $s_1 = s + p$, we have

$$\frac{ds_1}{s_1^2 + q} = a_2 \cdot dz. \tag{27.9}$$

Let us now consider all three possible cases: when $q = 0$, when $q > 0$, and when $q < 0$. When $q = 0$, then integrating both sides of (27.9), we get

$$-\frac{1}{s_1} = a_2 \cdot z + c,$$

for some integration constant c, thus $s_1(z) = -\dfrac{1}{a_2 \cdot z + c}$ and

$$s(z) = s_1(z) - p = -\frac{1}{a_2 \cdot z + c} - p.$$

This function is not everywhere defined—namely, it is not defined for $z = -c/a_2$, thus we will not consider it.

When $q > 0$, then for $s_2 = s_1/\sqrt{q}$, we have $s_1 = s_2 \cdot \sqrt{q}$ thus $ds_1 = \sqrt{q} \cdot ds_2$, $s_1^2 + q = q \cdot s_2^2 + q = q \cdot (s_2^1 + 1)$, and (27.9) becomes $\frac{1}{\sqrt{q}} \cdot \frac{ds_2}{s^2 + 1} = a_2 \cdot z$. Integrating leads to

$$\frac{1}{\sqrt{q}} \cdot \arctan(s_2) = a_2 \cdot z + c$$

hence $s_2(z) = \tan(\sqrt{q} \cdot a_2 \cdot z + \sqrt{q} \cdot c)$. This value is not always defined, thus $s_1(z) = \sqrt{q} \cdot s_2(z)$ and $s(z) = s_1(z) - p$ are also not always defined, so we will consider this case either.

For $q < 0$, for $s_2 = s_1/\sqrt{|q|}$, we similarly have $s_1 = s_2 \cdot \sqrt{|q|}$ thus $ds_1 = \sqrt{|q|} \cdot ds_2$, $s_1^2 + q = |q| \cdot s_2^2 + q = |q| \cdot (s_2^2 - 1)$, thus (27.9) becomes

$$\frac{1}{\sqrt{|q|}} \cdot \frac{ds_2}{s^2 - 1} = a_2 \cdot z. \tag{27.10}$$

One can easily check that

$$\frac{1}{s_2^2 - 1} = \frac{1}{2} \cdot \left(\frac{1}{s_2 - 1} - \frac{1}{s_2 + 1} \right),$$

thus integrating (27.10), we get

$$\frac{1}{2\sqrt{|q|}} \cdot (\ln(s_2 - 1) - \ln(s_2 + 1)) = a_2 \cdot z + c.$$

Multiplying both sides by $2\sqrt{|q|}$, we conclude that the difference

$$\ln(s_2 - 1) - \ln(s_2 + 1) = \ln\left(\frac{s_2 - 1}{s_2 + 1} \right)$$

is equal to a linear function z_1 of z. Thus, the fractional-linear ratio $\frac{s_2 - 1}{s_2 + 1}$ is equal to $\exp(z_1)$. The inverse to a fractional-linear transformation is also fractional linear, so $s_2(z)$ is a fractional-linear function of $\exp(z_1)$. The original function $s(z)$ is obtained from $s_2(z)$ by a linear transformation and is, thus, also a fractional-linear expression in terms of $\exp(z_1)$, i.e.,

$$s(z) = \frac{a \cdot \exp(z_1) + b}{c \cdot \exp(z_1) + 1}.$$

For this expression to always be defined, we need $c > 0$; else, if $c < 0$, it is not defined for $z_1 = -\ln(|c|)$. The expression for $s(z)$ can be written as

$$\frac{a}{c} + \text{const} \cdot \frac{1}{c \cdot \exp(z_1) + 1},$$

and for $z_2 = -z_1 + \ln(c)$, as

$$s(z) = \frac{a}{c} + \text{const} \cdot \frac{1}{1 + \exp(-z_2)}.$$

So, each such activation function $s(z)$ can be obtained if we:

• first, apply some linear transformation to z, getting z_2;
• then apply a sigmoid function; and
• finally, we apply a linear transformation to the result.

In the traditional neural network, as we mentioned earlier, we always apply some linear transformation *before* we apply the activation function, and we also apply some linear transformation *after* we apply the activation function. So, from the viewpoint of the above general formula of the traditional neural network, the class of functions which can be represented with K neurons by using the activation function $s(z)$ is exactly the same as the class of functions represented with K neurons by using the sigmoid function. In this sense, *sigmoid is the only shift-invariant activation function*—which explains its efficiency in traditional neural networks.

27.6   Limit Cases

In the previous section, we considered the generic case when $a_2 \neq 0$. To complete our analysis, we need to also consider the remaining case when $a_2 = 0$. This is a limit case of the generic case when $a_2 \to 0$. In this case, the formula (27.6) takes the following simplified form:

$$\frac{ds}{a_0 + a_1 \cdot s} = dz. \tag{27.11}$$

If $a_1 \neq 0$, then for $s_1 = a_0 + a_1 \cdot s$, we have $ds_1 = a_1 \cdot ds$, hence $ds = ds_1/a_1$, and (27.11) takes the form

$$\frac{1}{a_1} \cdot \frac{ds_1}{s_1} = dz.$$

Integrating, we get $\dfrac{1}{a_1} \cdot \ln(s_1) = z + c$, hence $\ln(s_1) = z_1 \stackrel{\text{def}}{=} a_1 \cdot z + a_1 \cdot c$, so $s_1(z) = \exp(z_1)$, and $s(z) = \dfrac{s_1(z) - a_0}{a_1} = \dfrac{1}{a_1} \cdot (\exp(z_1) - a_0)$. The resulting activation functions $s(z)$ can be obtained if we:

- first, apply some linear transformation to z, getting z_2;
- then apply an exponential function; and
- finally, we apply a linear transformation to the result.

Similarly to the generic case, we can thus conclude that the class of functions which can be represented with K neurons by the using the activation function $s(z)$ is exactly the same as the class of functions represented with K neurons by using the exponential function.

The only remaining case if $a_1 = 0$. In this case, (27.11) easily integrates into $\dfrac{s}{a_0} = z + c$, i.e, to a linear activation function $s(z) = a_0 \cdot z + a_0 \cdot c$. Mathematically, it is a legitimate case, but, of course, from the viewpoint of neural networks it makes no sense, since, as we have mentioned earlier, the whole point of activation functions is to cover *non*-linear functions.

27.7 We Need Multi-layer Neural Networks

The problem with traditional neural networks, as we mention in [4, 8, 9], is that they waste a lot of bits: for K neurons, any of $K!$ permutations results in exactly the same function. To decrease this duplication, we need to decrease the number of neurons K in each layer. So, instead of placing all nonlinear neurons in one layer, we place them in several consecutive layers. This is one of the main idea behind deep learning.

27.8 Which Activation Function Should We Use

In the first nonlinear layer, we make sure that a shift in the input—corresponding to a different estimate of the background noise—does not change the processing formula, i.e., that results $s(z + c)$ and $s(z)$ can be obtained from each other by applying an appropriate transformation—in this case, a fractional-linear transformation. We already know that this idea leads either to the sigmoid function (or to its limit case— exponential function).

So far, so good, but this logic does not work if we try to find out what activation function we should use in the *next* nonlinear layers. Indeed, the output of the first layer—which is the input to the second nonlinear layer—is *no longer shift*-invariant, it is invariant with respect to some more *complex* (fractional-linear) transformations. We know what to do when the input is shift-invariant, so a natural idea is to perform some *additional* transformation that will make the results shift-invariant. If we do that, then we will again be able to apply the sigmoid activation function, then again the additional transformation, etc.

These additional transformations should transform generic fractional-linear operations into shift. This means that the inverse of such a transformation should transform shifts into some fractional-linear operations. But this is exactly what we analyzed in the previous section—transformations that transform shifts into fractional-linear operations. We already know the formulas $s(z)$ for these transformations. In general, they are formed as follows:

- first, we apply some linear transformation to the input z, resulting in a linear combination $Z = p \cdot z + q$;
- then, we compute $Y = \exp(Z)$; and
- finally, we apply some fractional-linear transformation to the resulting value Y, getting y.

So, to get the inverse transformation, we need to reverse all three steps, starting with the last one:

- first, we apply a fractional-linear transformation to y, getting Y;
- then, we compute $Z = \ln(Y)$; and
- finally, we apply a linear transformation to Z, resulting in z.

27.9 This Leads Exactly to Squashing Functions

What happens if we first apply some sigmoid-type transformation moving us from shifts to tractional-linear operations and then an inverse-type transformation? The last step of the sigmoid-type transformation and the first step of the inverse-type transformation both apply fractional-linear transformations. Since the composition of fractional-linear transformations is fractional-linear, we can combine them into a single step. Thus, the resulting combined activation function can thus be described as follows:

- first, we apply some linear transformation L_1 to the input z, resulting in a linear combination $Z = L_1(z) = p \cdot z + q$;
- then, we compute $E = \exp(Z) = \exp(L_1(z))$;
- then, we apply some fractional-linear transformation F to $E = \exp(Z)$, getting $T = F(E) = F(\exp(L_1(z)))$;
- then, we compute $Y = \ln(T) = \ln(F(\exp(L_1(z))))$; and
- finally, we apply a linear transformation L_2 to Y, resulting in the final value $y = s(z) = L_2(Y) = L_2(\ln(F(\exp(L_1(z)))))$.

One can check that these are exactly squashing function! Thus, squashing functions can indeed be naturally explained by the invariance requirements.

27.10 Why Rectified Linear Functions

As an example of the above description, let us provide a family of squashing functions that tend to the rectified linear activation function $\max(z, 0)$. For this purpose, let us take:

- $L_1(z) = k \cdot z$, with $k > 0$, so that $E = \exp(L_1(z)) = \exp(k \cdot z)$;
- $F(E) = 1 + E$, so that $T = F(E) = \exp(k \cdot z) + 1$ and $Y = \ln(T) = \ln(\exp(k \cdot z) + 1)$; and
- $L_2(Y) = \dfrac{1}{k} \cdot Y$, so that the resulting activation function takes the form $s(z) = \dfrac{1}{k} \cdot \ln(\exp(k \cdot z) + 1)$.

Let us show that the above expression tends to the rectified linear activation function when $k \to \infty$.

When $z < 0$, then $\exp(k \cdot z) \to 0$, so $\exp(k \cdot z) + 1 \to 1$, $\ln(\exp(k \cdot z) + 1) \to 0$ and so $s(z) \to 0$.

On the other hand, when $z > 0$, then

$$\exp(k \cdot z) + 1 = \exp(k \cdot z) \cdot (1 + \exp(-k \cdot z)),$$

thus $\ln(\exp(k \cdot z) + 1) = k \cdot z + \ln(1 + \exp(-k \cdot z))$ and

$$s(z) = \frac{1}{k} \cdot \ln(\exp(k \cdot z) + 1) = z + \frac{1}{k} \cdot \ln(1 + \exp(-k \cdot z)).$$

When $k \to \infty$, we have $\exp(-k \cdot z) \to 0$, hence

$$1 + \exp(-k \cdot z) \to 1,$$

$\ln(1 + \exp(-k \cdot z)) \to 0$, so $\dfrac{1}{k} \cdot \ln(1 + \exp(-k \cdot z)) \to 0$ and indeed $s(z) \to z$.

References

1. Urenda, J.C., Csiszar, O., Csiszar, G., Dombi, J., Kosheleva, O., Kreinovich, V., Eigner, G.: Why squashing functions in multi-layer neural networks. In: Proceedings of the 2020 IEEE International Conference on Systems, Man, and Cybernetics SMC'2020, Toronto, Canada, pp. 296–300 (2020). Accessed from 11–14 Oct 2020
2. Goodfellow, I., Bengio, Y., Courville, A.: Deep Learning. MIT Press, Cambridge, Massachusetts (2016)
3. Fuentes, O., Parra, J., Anthony, E., Kreinovich, V.: Why rectified linear neurons are efficient: a possible theoretical explanations. In: Kosheleva, O., Shary, S., Xiang, G., Zapatrin, R. (eds.) Beyond Traditional Probabilistic Data Processing Techniques: Interval, Fuzzy, etc, pp. 603–613. Methods and Their Applications, Springer, Cham, Switzerland (2020)

4. Kreinovich, V., Kosheleva, O.: Deep learning (partly) demystified. In: Proceedings of the 2020 4th International Conference on Intelligent Systems, Metaheuristics & Swarm Intelligence ISMSI'2020, Thimpu, Bhutan (2020). Accessed from 18–19 Apr 2020
5. Csiszár, O., Csiszár, G., Dombi, J.: Interpretable Neural Networks Based on Continuous-Valued Logic and Multicriterion Decision Operators (2020). arXiv:1910.02486v2, posted on February 7, 2020
6. Dombi, J., Csiszár, O.: Operator-dependent modifiers in nilpotent logical systems. In: Proceedings of the 10th International Joint Conference on Computational Intelligence IJCCI'2018, Seville, Spain, pp. 126–134 (2018). Accessed from 18–20 Sept 2018
7. Dombi, J., Gera, Zs.: The approximation of piecewise linear membership functions and Lukasiewicz operators. Fuzzy Sets Syst. **154**, 275–286 (2005)
8. Baral, C., Fuentes, O., Kreinovich, V.: Why deep neural networks: a possible theoretical explanation. In: Ceberio, M., Kreinovich, V. (eds.) Constraint Programming and Decision Making: Theory and Applications, pp. 1–6. Springer, Berlin, Heidelberg (2018)
9. Kreinovich, V.: From traditional neural networks to deep learning: towards mathematical foundations of empirical successes. In: Shahbazova, S.N et al. (eds.), Proceedings of the World Conference on Soft Computing, Baku, Azerbaijan (2018). Accessed from 29–31 May 2018

Chapter 28
Data Processing: Neural Techniques, Part 2

Current artificial neural networks are very successful in many machine learning applications, but in some cases they still lag behind human abilities. To improve their performance, a natural idea is to simulate features of biological neurons which are not yet implemented in machine learning. One of such features is the fact that in biological neural networks, signals are represented by a train of spikes. Researchers have tried adding this spikiness to machine learning and indeed got very good results, especially when processing time series (and, more generally, spatio-temporal data). In this chapter, we provide a theoretical explanation for this empirical success.

Results from this chapter first appeared in [1].

28.1 Problem: Spiking Neural Networks

Why spiking neural networks: a historical reason. At this moment, artificial neural networks are the most successful—and the most promising—direction in Artificial Intelligence; see, e.g., [2].

Artificial neural networks are largely patterned after the way the actual biological neural networks work; see, e.g., [2, 3]. This patterning makes perfect sense: after all, our brains are the result of billions of years of improving evolution, so it is reasonable to conclude that many features of biological neural networks are close to optimal— not very efficient features would have been filtered out in this long evolutionary process.

However, there is an important difference between the current artificial neural networks and the biological neural networks:

- when some processing of the artificial neural networks is implemented in hardware—by using electronic or optical transformation—each numerical value is represented by the intensity (amplitude) of the corresponding signal;

J. C. Urenda and V. Kreinovich, *Algebraic Approach to Data Processing*,
Studies in Big Data 115, https://doi.org/10.1007/978-3-031-16780-5_28

- in contrast, in the biological neural networks, each value—e.g., the intensity of the sound or of the light—is represented by a series of instantaneous spikes, so that the original value is proportional to the frequency of these spikes.

Since simulating many other features of biological neural networks has led to many successes, a natural idea is to also try to emulate the spiking character of the biological neural networks.

Spiking neural networks are indeed efficient. Interestingly, adding spiking to artificial neural networks has indeed led to many successful applications, especially in processing temporal (and even spatio-temporal) signals; see, e.g., [4] and references therein.

But why? A biological explanation of the success of spiking neural networks—based on the above evolution arguments—makes perfect sense, but it would be nice to supplement it with a clear mathematical explanation—especially since, in spite of all the billions years of evolution, we humans are not perfect as biological beings, we need medicines, surgeries, and other artificial techniques to survive, and our brains often make mistakes.

What we do in this chapter. In this chapter, we consider the question of signal representation from the mathematical viewpoint, and we show that the spiking representation is indeed optimal in some reasonable sense.

28.2   Analysis of the Problem and the First Result

Looking for basic functions. In general, to represent a signal $x(t)$ means to approximate it as a linear combination of some basic functions. For example, it is reasonable to represent a periodic signal as a linear combination of sines and cosines. In more general cases—e.g., when analyzing weather—it makes sense to represent the observed values as a linear combination of functions t, t^2, etc., representing the trend and sines and cosines that describe the periodic part of the signal. To get a more accurate presentation, we need to take into account that the amplitudes of the periodic components can also change with time, so we end up with terms of the type $t \cdot \sin(\omega \cdot t)$.

If we analyze how radioactivity of a sample changes with time, a reasonable idea is to describe the measured values $x(t)$ as a linear combination of exponentially decaying functions $\exp(-k \cdot t)$ representing the decay of different isotopes, etc.

So, in precise terms, selecting a representation means selecting an appropriate family of basic functions. In general, we may have several parameters $c_1, \ldots, c_n$ characterizing functions from each family. Sometimes, there is only one parameter, as in sines and cosines. In other cases, we can have several parameters—e.g., in control applications, it makes sense to consider decaying periodic signals of the type $\exp(-k \cdot t) \cdot \sin(\omega \cdot t)$, with two parameters k and ω. In general, elements $b(t)$ of each such family can be described by a formula $b(t) = B(c_1, \ldots, c_n, t)$ corresponding to different tuples $c = (c_1, \ldots, c_n)$.

Dependence on parameters must be continuous in some reasonable sense. We want the dependence $B(c_1, \ldots, c_n, t)$ to be computable, and it is known that all computable functions are, in some reasonable sense, continuous; see, e.g., [5].

Indeed, in real life, we can only determine the values of all physical quantities c_i with some accuracy: measurements are always not 100% accurate, and computations always involve some rounding. For any given accuracy, we can provide the value with this accuracy—but it will practically never be the exact value. Thus, the approximate values of c_i are the only thing that our computing algorithm can use when computing the value $B(c_1, \ldots, c_n, t)$. This algorithm can ask for more and more accurate values of c_i, but at some point it must produce the result. At this point, we only known approximate values of c_i, i.e., we only know the interval of possible values of c_i. And for all the values of c_i from this interval, the result of the algorithm provides, with the given accuracy, the approximation to the desired value $B(c_1, \ldots, c_n, t)$. This is exactly what continuity is about!

One has to be careful here, since the real-life processes may actually be, for all practical purposes, discontinuous. Sudden collapses, explosions, fractures do happen.

For example, we want to make sure that a step-function which is equal to 0 for $t < 0$ and to 1 for $t \geq 0$ is close to an "almost" step function which is equal to 0 for $t < 0$, to 1 for $t \geq \varepsilon$ (for some small ε) and to t/ε for $t \in (0, \varepsilon)$.

In such situations, we cannot exactly describe the value at moment t—since the moment t is also measured approximately, but what we can describe is its values at a moment close to t. In other words, we can say that the two functions $a_1(t)$ and $a_2(t)$ are ε-close if:

- for every moment t_1, there exists moments t_{21} and t_{22} which are ε-close to t_1 (i.e., for which $|t_{2i} - t_1| \leq \varepsilon$) and for which $a_1(t_1)$ is ε-close to a convex combination of values $a_2(t_{2i})$, and
- for every moment t_2, there exists moments t_{11} and t_{12} which are ε-close to t_2 and for which $a_2(t_2)$ is ε-close to a convex combination of values $a_1(t_{1i})$.

Additional requirement. Since we consider linear combinations of basic functions, it does not make sense to have two basic functions that differ only by a constant: if $b_2(t) = C \cdot b_1(t)$, then there is no need to consider the function $b_2(t)$ at all; in each linear combination we can replace $b_2(t)$ with $C \cdot b_1(t)$.

We would like to have the simplest possible family of basic functions. How many parameters c_i do we need? The fewer parameters, the easier it is to adjust the values of these parameters, and the smaller the probability of *overfitting*—a known problem of machine learning in particular and of data analysis in general, when we fit the formula to the observed data and its random fluctuations too well and this make it much less useful in other cases where random fluctuations will be different.

We cannot have a family with no parameters at all—that would mean, in effect, that we have only one basic function $b(t)$ and we approximate every signal by an expression $C \cdot b(t)$ obtained by its scaling. This will be a very lousy approximation to real-life processes—since these processes are all different, they do not resemble each other at all.

So, we need at least one parameter. Since we are looking for the simplest possible family, we should therefore consider families depending on a single parameter c_1, i.e., families consisting of functions $b(t) = B(c_1, t)$ corresponding to different values of the parameter c_1.

Most observed processes are limited in time. From our viewpoint, we may view astronomical processes are going on forever—although, in reality, even they are limited by billions of years. However, in general, the vast majority of processes that we observe and that we want to predict are limited in time: a thunderstorm stops, a hurricane end, after-shocks of an earthquake stop, etc.

From this viewpoint, to get a reasonable description of such processes, it is desirable to have basic functions which are also limited in time, i.e., which are equal to 0 outside some finite time interval. This need for finite duration is one of the main reasons in many practical problems, a decomposition into wavelets performs much better that a more traditional Fourier expansion into linear combinations of sines and cosines; see, e.g., [6] and references therein.

Shift- and scale-invariance. Processes can start at any moment of time. Suppose that we have a process starting at moment 0 which is described by a function $x(t)$. What if we start the same process t_0 moments earlier? At each moment t, the new process has been happening for the time period $t + t_0$. Thus, at the moment t, the new process is at the same stage as the original process will be at the future moment $t + t_0$. So, the value $x'(t)$ of a quantity characterizing the new process is equal to the value $x(t + t_0)$ of the original process at the future moment of time $t + t_0$.

There is no special starting point, so it is reasonable to require that the class of basic function not change if we simply change the starting point. In other words, we require that for every t_0, the shifted family $\{B(c_1, t + t_0)\}_{c_1}$ coincides with the original family $\{B(c_1, t)\}_{c_1}$.

Similarly, processes can have different speed. Some processes are slow, some are faster. If a process starting at 0 is described by a function $x(t)$, then a λ times faster process is characterized by the function $x'(t) = x(\lambda \cdot t)$. There is no special speed, so it is reasonable to require that the class of basic function not change if we simply change the process's speed. In other words, we require that for every $\lambda > 0$, the "scaled" family $\{B(c_1, \lambda \cdot t)\}_{c_1}$ coincides with the original family $\{B(c_1, t)\}_{c_1}$.

Now, we are ready for the formal definitions.

Definition 28.1 We say that a function $b(t)$ is *limited in time* if it equal to 0 outside some interval.

Definition 28.2 We say that a function $b(t)$ is a *spike* if it is different from 0 only for a single value t. This non-zero value is called the *height* of the spike.

Definition 28.3 Let $\varepsilon > 0$ be a real number. We say that the numbers a_1 and a_2 are *ε-close* if $|a_1 - a_2| \le \varepsilon$.

Definition 28.4 We say that the functions $a_1(t)$ and $a_2(t)$ are ε-*close* if:

- for every moment t_1, there exists moments t_{21} and t_{22} which are ε-close to t_1 (i.e., for which $|t_{2i} - t_1| \leq \varepsilon$) and for which $a_1(t_1)$ is ε-close to a convex combination of values $a_2(t_{2i})$, and
- for every moment t_2, there exists moments t_{11} and t_{12} which are ε-close to t_2 and for which $a_2(t_2)$ is ε-close to a convex combination of values $a_1(t_{1i})$.

Comment. One can check that this definition is equivalent to the inequality $d_H(A_1, A_2) \leq \varepsilon$ bounding the Hausdorff distance $d_H(A_1, A_2)$ between the two sets A_i each of which is obtained from the closure C_i of the graphs of the corresponding function $a_i(t)$ by adding the whole vertical interval $t \times [a, b]$ for every two points (t, a) and (t, b) with the same first coordinate from the closure C_i.

Definition 28.5 We say that a mapping $B(c_1, t)$ that assigns, to each real number c_1, a function $b(t) = B(c_1, t)$ is *continuous* if, for every value c_1 and for every $\varepsilon > 0$, there exists a real number $\delta > 0$ such that, if c_1' is δ-close to c_1, then the function $b(t) = B(c_1, t)$ is ε-close to the function $b'(t) = B(c_1', t)$.

Definition 28.6 By a *family of basic functions*, we mean a continuous mapping for which:

- for each c_1, the function $b(t) = B(c_1, t)$ is limited in time, and
- if c_1 and c_1' are two different numbers, then the functions $b(t) = B(c_1, t)$ and $b'(t) = B(c_1', t)$ cannot be obtained from each other by multiplication by a constant.

Definition 28.7 We say that a family of basic functions $B(c_1, t)$ is *shift-invariant* if for each t_0, the following two classes of functions of one variable coincide:

$$\{B(c_1, t)\}_{c_1} = \{B(c_1, t + t_0)\}_{c_1}.$$

Definition 28.8 We say that a family of basic functions $B(c_1, t)$ is *scale-invariant* if for each $\lambda > 0$, the following two classes of functions of one variable coincide:

$$\{B(c_1, t)\}_{c_1} = \{B(c_1, \lambda \cdot t)\}_{c_1}.$$

Proposition 28.1 *If a family of basic functions $B(c_1, t)$ is shift- and scale-invariant, then for every c_1, the corresponding function $b(t) = B(c_1, t)$ is a spike, and all these spikes have the same height.*

Discussion. This result explains the efficiency of spikes: namely, a family of spikes is the only one which satisfies the reasonable conditions of shift- and scale-invariance, i.e., the only family that does not change if we change the starting point of the process and/or change the process's speed.

Proof Let us assume that the family of basic functions $B(c_1, t)$ is shift- and scale-invariant. Let us prove that all the functions $b(t) = B(c_1, t)$ are spikes.

$1°$. First, we prove that none of the functions $B(c_1, t)$ is identically 0.

Indeed, the zero function can be contained from any other function by multiplying that other function by 0—and this would violate the second part of Definition 28.6 (of a family of basic functions).

$2°$. Let us prove that each function from the given family is a spike.

Indeed, each of the functions $b(t) = B(c_1, t)$ is not identically zero, i.e., it attains non-zero values for some t. By the Definition 28.6 of a family of basic functions, each of these functions is limited in time, i.e., the values t for which the function $b(t)$ is non-zero are bounded by some interval. Thus, the values $t_- \overset{\text{def}}{=} \inf\{t : b(t) \neq 0\}$ and $t_+ \overset{\text{def}}{=} \sup\{t : b(t) \neq 0\}$ are finite, with $t_- \leq t_+$.

Let us prove that we cannot have $t_- < t_+$. Indeed, in this case, the interval $[t_-, t_+]$ is non-degenerate. Thus, by an appropriate combination of shift and scaling, we will be able to get this interval from any other non-degenerate interval $[a, b]$, with $a < b$: indeed, it is sufficient to take the transformation $t \to \lambda \cdot t + t_0$, where $\lambda = \dfrac{t_+ - t_-}{b - a}$ and $t_0 = \lambda \cdot a - t_-$. For each of these transformations, due to shift- and scale-invariance of the family, the correspondingly re-scaled function $b'(t) = b(\lambda \cdot t + t_0)$ also belongs to the family $B(c_1, t)$, and for this function, the corresponding values t'_- and t'_+ will coincide with a and b. All these functions are different—so, we will have a 2-dimensional family of functions (i.e., a family depending on 2 parameters), which contradicts to our assumption that the family $B(c_1, t)$ is one-dimensional.

The fact that we cannot have $t_- < t_+$ means that we should have $t_- = t_+$, i.e., that every function $b(t)$ from our family is indeed a spike.

$3°$. To complete the proof, we need to prove that all the spikes that form the family $B(c_1, t)$ have the same height.

Let us describe this property in precise terms. Let $b_1(t)$ and $b_2(t)$ be any two functions from the family. According to Part 2 of this proof, both functions are spikes, so:

- the value $b_1(t)$ is only different from 0 for some value t_1; let us denote the corresponding height $b_1(t_1)$ by h_1;
- similarly, the value $b_2(t)$ is only different from 0 for some value t_2; let us denote the corresponding height $b_2(t_2)$ by h_2.

We want to prove that $h_1 = h_2$.

Indeed, since the function $b_1(t)$ belongs to the family, and the family is shift-invariant, then for $t_0 \overset{\text{def}}{=} t_1 - t_2$, the shifted function $b'_1(t) \overset{\text{def}}{=} b_1(t + t_0)$ also belongs to this family. The shifted function is non-zero when $t + t_0 = t_1$, i.e., when $t = t_1 - t_0 = t_2$, and it has the same height h_1.

If $h_1 \neq h_2$, this would contradict to the second part of Definition 28.6 (of the family of basic functions)—because then we would have two functions $b'_1(t)$ and

$b_2(t)$ in this family, which can be obtained from each other by multiplying by a constant. Thus, the heights must be the same.

The proposition is proven.

28.3 Main Result: Spikes Are, in Some Reasonable Sense, Optimal

It is desirable to check whether spiked neurons are optimal. In the previous section, we showed that spikes naturally appear if we require reasonable properties like shift- and scale-invariance. This provides some justification for the spiked neural networks.

However, the ultimate goal of neural networks is to solve practical problems. From this viewpoint, we need to take into account that a practitioner is not interested in invariance or other mathematical properties, a practitioner wants to optimize some objective function. So, from the practitioner's viewpoint, the main question is: are spiked neurons optimal?

Different practitioners have different optimality criteria. The problem is that, in general, different practitioners may have different optimality criteria. In principle, we can pick one such criterion (or two or three) and analyze which families of basic functions are optimal with respect to these particular criterion—but this will not be very convincing to a practitioner who has a different optimality criterion.

An ideal explanation should work for *all* reasonable optimality criteria. This is what we aim at in this section. To achieve this goal, let us analyze what we mean by an optimality criterion, and which optimality criteria can be considered reasonable. In this analysis, we will follow a general analysis of computing-related optimization problems performed in Chap. 3 (see also [7]).

Definition 28.9 For each family of basic functions $B(c_1, t)$ and for each value t_0, by its *shift* $T_{t_0}(B)$, we mean a family that assigns, to each number c_1, a function $B(c_1, t + t_0)$.

Definition 28.10 We say that an optimality criterion on the class of all families of basic functions is *shift-invariant* if for every two families B and B' and for each t_0, $B \leq B'$ implies that $T_{t_0}(B) \leq T_{t_0}(B')$.

Definition 28.11 For each family of basic functions $B(c_1, t)$ and for each value $\lambda > 0$, by its *scaling* $S_\lambda(B)$, we mean a family that assigns, to each number c_1, a function $B(c_1, \lambda \cdot t)$.

Definition 28.12 We say that an optimality criterion on the class of all families of basic functions is *scale-invariant* if for every two families B and B' and for each $\lambda > 0$, $B \leq B'$ implies that $S_\lambda(B) \leq S_\lambda(B')$.

Now, we are ready to formulate our main result.

Proposition 28.2 *For every final shift- and scale-invariant optimality criterion on the class of all families of basic functions, all elements of the optimal family are spikes of the same height.*

Proof From the corresponding result from Chap. 3, it follows that the optimal family B_{opt} is itself shift- and scale-invariant. Thus, the desired result follows from Proposition 28.1.

References

1. Beer, M., Urenda, J., Kosheleva, O., Kreinovich, V.: Why spiking neural networks are efficient: a theorem. In: Proceedings of the 18th International Conference on Information Processing and Management of Uncertainty in Knowledge-Based Systems IPMU'2020. Lisbon, Portugal, 15–19 June 2020, pp. 59–69
2. Goodfellow, I., Bengio, Y., Courville, A.: Deep Learning. MIT Press, Cambridge, Massachusetts (2016)
3. Bishop, C.M.: Pattern Recognition and Machine Learning. Springer, New York (2006)
4. Kasabov, N.K. (ed.): Time-Space. Spiking Neural Networks and Brain-Inspired Artificial Intelligence, Springer, Cham, Switzerland (2019)
5. Weihrauch, K.: Computable Analysis: An Introduction. Springer, Berlin, Heidelberg, New York (2000)
6. Addison, P.S.: The Illustrated Wavelet Transform Handbook: Introductory Theory and Applications in Science, Engineering, Medicine and Finance. CRC Press, Boca Raton, Florida (2016)
7. Nguyen, H.T., Kreinovich, V.: Applications of Continuous Mathematics to Computer Science. Kluwer, Dordrecht (1997)

Chapter 29
Data Processing: Fuzzy Techniques, Part 1

29.1 Why Fuzzy Techniques

One of the challenges of neural techniques such as deep learning is that these technique are a black box, its results do not come with any explanations. If we could add some natural-language explanations, that would make these results more convincing and thus, more acceptable. Thus, we need explanations, we need understandability of the results.

Understandability means that we should be able to describe the computations by using words from natural language. One of the main challenges in coming up with such a description is that natural language is imprecise (fuzzy), so it is difficult to find the relation between imprecise words from natural language and precise algorithms. In solving this challenge, it is natural to use the experience of researchers who came up with such a relationship from the other side of it: by trying to translate natural-language knowledge into precise terms.

29.2 Fuzzy Techniques: Main Ideas

This experience led to the design on fuzzy logic by Lotfi Zadeh; see, e.g., [1–6]. Lotfi Zadeh, a specialist in control and an author of a successful textbook on control, noticed, in the early 1960s, a puzzling phenomenon: that human-led control often leads to much better results than even the optimal automatic control. The answer to this puzzle was clear: humans use additional knowledge which was not taken into account when the automatic controllers were designed. The reason why this additional knowledge was not taken into account is that this knowledge is not described in precise terms, it is described by using imprecise words from natural language. For example, an operator may say: if the pressure drops a little bit, increase a little bit the flow of the chemical into the chamber; here, "a little bit" does not have a precise

© The Author(s), under exclusive license to Springer Nature Switzerland AG 2022
J. C. Urenda and V. Kreinovich, *Algebraic Approach to Data Processing*,
Studies in Big Data 115, https://doi.org/10.1007/978-3-031-16780-5_29

meaning. Zadeh invented a methodology for translating this "fuzzy" knowledge into precise terms, a methodology that he called *fuzzy logic*, or, more generally, *fuzzy techniques*.

His main point is that in contrast to exact statements like "pressure is below 1.2 atmospheres"—which is always either true or false—about the statements that include natural-language words—like "the drop from 1.3 to 1.2 means that the pressure dropped a little bit"—experts are not sure. The smaller the drop, the larger the expert's degree of confidence that this statement is true. For each value of the corresponding quantity (e.g., pressure), we can gauge the expert's degree of confidence in the corresponding statement by asking the expert to mark it on a scale, e.g., from 0 to 10. The resulting mark depends on what scale we use: from 0 to 5 or from 0 to 10 or form 0 to any other number. To make these estimates uniform, a reasonable idea is to divide the mark by the largest number on the scale, so that, e.g., 7 on a scale from 0 to 10 becomes $7/10 = 0.7$. In this new scale, 1 means that the expert is absolutely confident that this statement is true, 0 means that the expert is absolutely confident that the statement is false, and values between 0 and 1 correspond to intermediate degrees of confidence.

29.3 Fuzzy Techniques: Logic

The reason why this methodology is called fuzzy *logic* is that in addition to simple statements—like the ones above—expert knowledge often contains statements that include *logical connectives* like "and" and "or". For example, an expert can recommend a certain action if the pressure dropped a little bit *and* the temperature increased somewhat. How can we gauge our degree of certainty in such composite statements? It would be great if we could similarly ask the expert to estimate his/her degree of confidence for all possible pairs of values (pressure, temperature). If we have a composite statement combining three or four different statements, we would need to consider all possible triples or quadruples. Even if we consider a reasonable number 20-30 of possible values of each quantity, it makes sense to ask the expert about all 30 values, but asking about all $30^4 = 810000$ possible quadruples is not realistic. Since we cannot directly elicit the degree of confidence in all such composite statements directly from the expert, we need to be able to estimate this degree based on whatever information we can elicit—i.e., based on the expert's degrees of confidence in the component statements.

In precise terms, we need a procedure that would take, as input, the degrees of confidence a and b in two statements A and B and return an estimate for the expert's degree of confidence in a composite statement $A \& B$. We will denote this estimate by $f_\&(a, b)$. The corresponding function $f_\&$ is known as an *"and"-operation*, or, for historical reason, a *t-norm*.

Since the statements "A and B" and "B and A" mean the same thing, it is reasonable to require that for these two statements, we have the same degree of confidence, i.e., that $f_\&(a, b) = f_\&(b, a)$. In other words, an "and"-operation must be commutative.

When A is false, clearly $A \& B$ is false too, so we must have $f_\&(0, b) = 0$ for all b. When A is true, our degree of confidence in $A \& B$ is the same as our degree of confidence in B, i.e., we must have $f_\&(1, b) = b$.

Similarly, we need a procedure that would take, as input, the degrees of confidence a and b in two statements A and B and return an estimate for the expert's degree of confidence in a composite statement $A \vee B$. We will denote this estimate by $f_\vee(a, b)$. The corresponding function $f_\vee$ is known as an *"or"-operation*, or, for historical reason, a *t-conorm*.

Since the statements "A or B" and "B or A" mean the same thing, it is reasonable to require that for these two statements, we have the same degree of confidence, i.e., that $f_\vee(a, b) = f_\vee(b, a)$. In other words, an "or"-operation must be commutative.

When A is true, clearly $A \vee B$ is true too, so we must have $f_\vee(1, b) = 1$ for all b. When A is false, our degree of confidence in $A \vee B$ is the same as our degree of confidence in B, i.e., we must have $f_\vee(0, b) = b$.

References

1. Belohlavek, R., Dauben, J.W., Klir, G.J.: Fuzzy Logic and Mathematics: A Historical Perspective. Oxford University Press, New York (2017)
2. Klir, G., Yuan, B.: Fuzzy Sets and Fuzzy Logic. Prentice Hall, Upper Saddle River, New Jersey (1995)
3. Mendel, J.M.: Uncertain Rule-Based Fuzzy Systems: Introduction and New Directions. Springer, Cham, Switzerland (2017)
4. Nguyen, H.T., Walker, C.L., Walker, E.A.: A First Course in Fuzzy Logic. Chapman and Hall/CRC, Boca Raton, Florida (2019)
5. Novák, V., Perfilieva, I., Močkoř, J.: Mathematical Principles of Fuzzy Logic. Kluwer, Boston, Dordrecht (1999)
6. Zadeh, L.A.: Fuzzy sets. Inf. Control **8**, 338–353 (1965)

Chapter 30
Data Processing: Neural and Fuzzy Techniques

In the previous chapters, we argued that to make neural techniques more understandable, we need to use fuzzy techniques. In this chapter, we show that this use leads to one more explanation of why rectified linear neurons are so successful.

Results of this chapter first appeared in [1].

30.1 Problem: Computations Should Be Fast and Understandable

Two important challenges of data processing: computation speed and understandability. In most practical problems, we need to process a large amount of data—and we need to make a decision reasonably fast:

- if we predict weather, we need to take into account all the results of today's measurements of temperature, wind speed and direction, etc., in a given geographic areas, satellite images, historical data—and get the prediction of tomorrow's weather the same day: otherwise, our prediction will be useless;
- if we decide whether to give a person a loan, we need to take into account this person's financial history, financial history of similar customers, general economic situation in the region, etc.—and get the result fast, otherwise the customer may lose the business opportunity for which he/she is seeking this loan.

So, we need all the computations to be as fast as possible.

We also ideally want the computations to be understandable.

- When a weatherperson on the TV predict's tomorrow's weather, it is much more convincing if this person explains why we should expect strong winds, or, vice versa, perfect weather. These explanations may not be quantitative, usually, qualitative explanations are good enough.

J. C. Urenda and V. Kreinovich, *Algebraic Approach to Data Processing*, Studies in Big Data 115, https://doi.org/10.1007/978-3-031-16780-5_30

- When we explain, to the person, why he/she is not getting a loan while his/her friends are, we need to have some reasonable explanations—at least to avoid lawsuits claiming gender-based, age-based, or race-based bias.

How can we achieve these two goals?

These two goals lead to neural and fuzzy. As we have shown, a natural way to speed up computations is to use neural techniques, and a natural way to make computations understandable—i.e., to relate them to natural-language explanations—is to use fuzzy techniques.

We want our computations to be both fast and understandable. Understandable means that we have to use some "and"- and "or"-operations. We thus want these operations to be fast. The fastest possible computations are computations on a 1-layer neural network, in which thus "and"-operation is computed by a single neuron, and in which the "or"-operation can also be computed by a single neuron. So, natural questions are:

- which "and"- and "or"-operations can be computed by a 1-layer neural network, and
- what activation functions allow computing "and"- and "or"-operations by such neural networks.

What we do in this chapter. In this chapter, we provide answers to both questions, namely:

- we show that the only "and"- and "or"-operations which can be computed by a 1-layer neural network are $\max(0, a + b - 1)$ and $\min(a + b, 1)$, and
- we show that the only activation function allowing such fast computations are equivalent to *rectified linear neurons*—which probably provides some explanations for the current success of such activation functions.

We also show that if we allow linear pre-processing after a single neuron, then we also represent $\min(a, b)$ and $\max(a, b)$. If we allow several neurons in a 2-layer network, then, in effect, we can compute any "and"- and "or"-operations.

30.2   Definitions and the Main Results

Definition 30.1 By an *"and"-operation*, we mean a function

$$f_\& : [0, 1] \times [0, 1] \to [0, 1]$$

for which the following properties are satisfied:

- $f_\&(a, b) = f_\&(b, a)$ for all a and b,
- $f_\&(0, b) = 0$ and $f_\&(1, b) = b$ for all b.

Definition 30.2 By an *"or"-operation*, we mean a function

$$f_\vee : [0, 1] \times [0, 1] \to [0, 1]$$

for which the following properties are satisfied:

- $f_\vee(a, b) = f_\vee(b, a)$ for all a and b,
- $f_\vee(0, b) = b$ and $f_\vee(1, b) = 1$ for all b.

Comment. Usually, for both "and"- and "or"-operations, other properties are required as well—namely, continuity, monotonicity, and associativity—but for our main results, we do not need these additional properties.

Definition 30.3 We say that a function $f(x_1, \ldots, x_n)$ *can be represented by a 1-layer neural network* if this function can be represented in the form

$$f(x_1, \ldots, x_n) = s(w_0 + w_1 \cdot x_1 + \cdots + w_n \cdot x_n)$$

for some function $s(z)$ and for some values w_i. The corresponding function $s(z)$ is called an *activation function*.

Definition 30.4 By a *rectified linear* function, we mean a function

$$s_0(z) = \max(0, z).$$

Definition 30.5 We say that two activation functions $s_1(z)$ and $s_2(z)$ are *equivalent* if for some constants a_{ij} and b_{ij}, we have

$$s_1(z) = a_{10} + a_{12} \cdot s_2(b_{10} + b_{11} \cdot z) + a_{1z} \cdot z$$

and

$$s_2(z) = a_{20} + a_{21} \cdot s_1(b_{20} + b_{21} \cdot z) + a_{2z} \cdot z$$

for all z.

Comment. This way, the corresponding multi-layer neural networks represent, in effect, the same class of functions, since each non-linear layer is equivalent to adding extra linear transformations before and after the non-linear layer representing another activation function.

Proposition 30.1 *The only "and"-operation that can be represented by a 1-layer neural network is* $\max(0, a + b - 1)$, *and all activation functions allowing such a representation are equivalent to the rectified linear function.*

Proposition 30.2 *The only "or"-operation that can be represented by a 1-layer neural network is* $\min(a + b, 1)$, *and all activation functions allowing such a representation are equivalent to the rectified linear function.*

Comment. These results provide another explanation for why rectified linear activation functions are so successful in deep neural networks.

Proof of Proposition 30.1 Let us consider an "and"-operation $f_\&(a, b)$ which can be represented by a 1-layer neural network. By definition of such a representation, this means that $f_\&(a, b) = s(w_0 + w_a \cdot a + w_b \cdot b)$ for some function $s(z)$ and for some coefficients w_i.

By definition of an "and"-operation, we have $f_\&(a, b) = f_\&(b, a)$ for all a and b. Thus, the expression $s(w_0 + w_a \cdot a + w_b \cdot b)$ should not change if we swap a and b: $s(w_0 + w_a \cdot a + w_b \cdot b) = s(w_0 + w_a \cdot b + w_b \cdot a)$. Therefore, we must have $w_a = w_b$, i.e., $f_\&(a, b) = s(w_0 + w_a \cdot a + w_a \cdot b)$, and thus,

$$f_\&(a, b) = s(w_0 + w_a \cdot (a + b)). \tag{30.1}$$

Let us introduce an auxiliary function $t(z) \stackrel{\text{def}}{=} s(w_0 + w_a \cdot z)$. This function is, by the definition of equivalence, equivalent to $s(z)$. In terms of this auxiliary function, the formula (30.1) takes the following simplified form:

$$f_\&(a, b) = t(a + b). \tag{30.2}$$

For $a = 0$, by definition of an "and"-operation, we have $f_\&(0, b) = 0$ for all $b \in [0, 1]$, thus $t(z) = 0$ for all $z \in [0, 1]$.

For $a = 1$, by definition of an "and"-operation, we have $f_\&(1, b) = b$ for all $b \in [0, 1]$, thus $t(1 + b) = b$ for all $b \in [0, 1]$. For $z = 1 + b$, we have $z \in [1, 2]$ and $b = z - 1$, thus $t(z) = z - 1$ for all $z \in [1, 2]$. So, we have:

- $t(z) = 0$ for $z \in [0, 1]$, and
- $t(z) = z - 1$ for $z \in [1, 2]$.

These two cases can be combined into a single formula

$$t(z) = \max(0, z - 1). \tag{30.3}$$

Substituting this expression for $t(z)$ into the formula (30.2), we conclude that $f_\&(a, b) = \max(0, a + b - 1)$. So, this "and"-operation is indeed the only one that can be represented by a 1-layer neural network.

Which activation functions can be used for this representation? From the formula (30.3), we can see that $t(z)$ is indeed equivalent to the rectified linear activation function. Since the original function $s(z)$ is equivalent to $t(z)$, we can conclude that $s(z)$ is also equivalent to the rectified linear activation function. Thus, the 1-layer representation of an "and"-operation is only possible if we use rectified linear neurons.

The proposition is proven.

Proof of Proposition 30.2 Let us now consider an "or"-operation $f_\vee(a, b)$ which can be represented by a 1-layer neural network. By definition of such a representation, this means that $f_\vee(a, b) = s(w_0 + w_a \cdot a + w_b \cdot b)$ for some function $s(z)$ and for some coefficients w_i.

By definition of an "or"-operation, we have $f_\vee(a, b) = f_\vee(b, a)$ for all a and b. Thus, the expression $s(w_0 + w_a \cdot a + w_b \cdot b)$ should not change if we swap a and b: $s(w_0 + w_a \cdot a + w_b \cdot b) = s(w_0 + w_a \cdot b + w_b \cdot a)$. Therefore, we must have $w_a = w_b$, i.e., $f_\vee(a, b) = s(w_0 + w_a \cdot a + w_a \cdot b)$, and thus,

$$f_\vee(a, b) = s(w_0 + w_a \cdot (a + b)). \tag{30.4}$$

Similar to the Proof of Proposition 30.1, let us introduce an auxiliary function $t(z) \overset{\text{def}}{=} s(w_0 + w_a \cdot z)$. This function is, by the definition of equivalence, equivalent to $s(z)$. In terms of this auxiliary function, the formula (30.4) takes the following simplified form:

$$f_\vee(a, b) = t(a + b). \tag{30.5}$$

For $a = 0$, by definition of an "or"-operation, we have $f_\vee(0, b) = b$ for all $b \in [0, 1]$, thus $t(z) = z$ for all $z \in [0, 1]$.

For $a = 1$, by definition of an "or"-operation, we have $f_\&(1, b) = 1$ for all $b \in [0, 1]$, thus $t(1 + b) = 1$ for all $b \in [0, 1]$. For $z = 1 + b$, we have $z \in [1, 2]$ and $b = z - 1$, thus $t(z) = 1$ for all $z \in [1, 2]$. So, we have:

- $t(z) = z$ for $z \in [0, 1]$, and
- $t(z) = 1$ for $z \in [1, 2]$.

These two cases can be combined into a single formula

$$t(z) = \min(z, 1). \tag{30.6}$$

Substituting this expression for $t(z)$ into the formula (30.5), we conclude that $f_\vee(a, b) = \min(1, a + b)$. So, this "or"-operation is indeed the only one that can be represented by a 1-layer neural network.

Which activation functions can be used for this representation? One can easily see that the expression (30.6) can be represented in an equivalent form $t(z) = 1 - \max(1 - z, 0)$, so $t(z)$ is indeed equivalent to the rectified linear activation function. Since the original function $s(z)$ is equivalent to $t(z)$, we can conclude that $s(z)$ is also equivalent to the rectified linear activation function. Thus, the 1-layer representation of an "or"-operation is only possible if we use rectified linear neurons.

The proposition is proven.

30.3 Auxiliary Result: What Can We Do with Two-Layer Networks

What about other "and"- and "or"-operations? In this chapter, we have shown that only the operations $f_\&(a, b) = \max(0, a + b - 1)$ and $f_\vee(a, b) = \min(a + b, 1)$ can be represented by 1-layer neural networks. How many layers do we need to represent general "and"- and "or"-operations?

It is known—see, e.g., [2]—that for every continuous "and"- (or "or"-) operation $f(a, b)$ and for every $\varepsilon > 0$, then exists a function $F(z)$ for which an "and"- (or, respectively, "or"-) operation

$$g(a, b) = F^{-1}(F(a) + F(b)) \tag{30.7}$$

satisfies the property $|f(a, b) - g(a, b)| \le \varepsilon$ for all a and b. (Of course, for this result to be true, it is not sufficient to have the above simplified definitions of "and"- and "or"-operations: we also need to assume associativity and monotonicity.)

For very small ε, the operations $f(a, b)$ and $g(a, b)$ are practically indistinguishable. So, from practical viewpoint, every "and"-operation and every "or"-operation can be represented in the form (30.7). Every function of this form can be computed by a 2-layer neural network:

- in the first layer, we use the inputs a and b to compute the values $a' = F(a)$ and $b' = F(b)$;
- then, in the second layer, we compute the value $F^{-1}(a' + b')$, which is exactly the desired value $F^{-1}(F(a) + F(b))$.

So, from the practical viewpoint, every "and"-operation and every "or"-operation can be computed by a 2-layer neural network.

For example, a widely used "and"-operation $f_\&(a, b) = a \cdot b$ can be computed as $\exp(\ln(a) + \ln(b))$, with $F(z) = \ln(z)$ and the inverse function $F^{-1}(z) = \exp(z)$. Similarly, a widely used "or"-operation $f_\vee(a, b) = a + b - a \cdot b$ can be computed in the form (7) with $F(z) = \ln(1 - z)$ and $F^{-1}(z) = 1 - \exp(z)$.

When is it sufficient to have a single neuron with linear post-processing? We have shown that, from the practical viewpoint, all "and"- and "or"-operations can be represented by a 2-layer neural network. Interestingly, some "and"- and "or"-operations $f(a, b)$ can be represented by a single neuron if we allow an additional linear post-processing. For example, one can easily see that $\min(a, b) = b - \max(0, b - a)$ and $\max(a, b) = a + \max(0, b - a)$.

It turns out that these are the only "and"- and "or"-operations which can be thus represented.

Definition 30.6 We say that a continuous monotonic associative "and"-operation $f_\&(a, b)$ *can be computed by a single neuron with linear post-processing* if we have

$$f_\&(a, b) = c_0 + c_a \cdot a + c_b \cdot b + s(w_0 + w_a \cdot a + w_b \cdot b). \tag{30.8}$$

Definition 30.7 We say that a continuous monotonic associative "or"-operation $f_\vee(a, b)$ *can be computed by a single neuron with linear post-processing if we have*

$$f_\vee(a, b) = c_0 + c_a \cdot a + c_b \cdot b + s(w_0 + w_a \cdot a + w_b \cdot b). \qquad (30.9)$$

Proposition 30.3 *The only "and"-operations that can be computed by a single neuron with linear post-processing are* $\max(0, a + b - 1)$ *and* $\min(a, b)$. *All activation functions allowing such a computation are equivalent to the rectified linear function.*

Proposition 30.4 *The only "or"-operations that can be computed by a single neuron with linear post-processing are* $\min(a + b, 1)$ *and* $\max(a, b)$. *All activation functions allowing such a computation are equivalent to the rectified linear function.*

Proof of Propositions 30.3 and 30.4 First of all, let us somewhat simplify the expressions (30.8) and (30.9) for the corresponding operation $f(a, b)$.

We cannot have $w_a = w_b = 0$ because then, the function $f(a, b)$ would be linear, and it is easy to show that no linear function can satisfy all the requirements of an "and"-operation or of an "or"-operation. Thus, either $w_a \neq 0$ or $w_b \neq 0$ (or both).

If $w_a = 0$, then, due to commutativity of $f(a, b)$, we can swap a and b and get an expression with $w_a \neq 0$. Thus, without losing generality, we can assume that $w_a \neq 0$.

We can thus introduce an auxiliary function $t(z) = c_0 + s(w_0 + w_a \cdot z)$. In terms of this auxiliary function, formulas (30.8) and (30.9) take the form

$$f(a, b) = c_a \cdot a + c_b \cdot b + t(a + k \cdot b), \qquad (30.10)$$

where $k \stackrel{\text{def}}{=} w_b/w_a$.

If $k = 1$, then the expression $t(a + k \cdot b)$ is symmetric with respect to a and b. Since for both types of operations, the function $f(a, b)$ is commutative, we thus conclude that the difference

$$c_a \cdot a + c_b \cdot b = f(a, b) - t(a + b)$$

is also commutative. Therefore, $c_a = c_b$, hence the whole expression (30.10) depends only on the sum $a + b$, i.e., has the form $F(a + b)$ for some function $F(z)$. This means that each such function is computable by a 1-layer neural network, and all "and"- and "or"-operations which can be thus represented have been described in Propositions 1 and 2.

To complete the proof, it is therefore necessary to consider the case when $k \neq 1$, i.e., when the lines $a + k \cdot b = \text{const}$ are not parallel to the diagonal $a = b$ of the square $[0, 1] \times [0, 1]$. Each line $a + k \cdot b = \text{const}$ intersects the borderline of the square at two points. On the borderline—i.e., when one of the values a and b is equal to 0 or to 1—the value of an "and"- or "or"-operation is uniquely determined by the corresponding Definition (Definition 30.1 or Definition 30.2). Since the function $f(a, b)$ is linear on this line, its values for all the points from this line are uniquely determined by the values at these two borderline points. Thus, for each k, we uniquely determine all the values $f(a, b)$ for all the pairs (a, b).

One can check that the only case when the resulting function is commutative and associative is the case $k = -1$, in which case we indeed get $\min(a, b)$ and $\max(a, b)$. We can also easily check that in both case, the activation function $t(z)$ is indeed equivalent to the rectified linear function. The propositions are proven.

Remaining open problems. It is known (see, e.g., [3]) that functions represented as linear combinations of the results of 1-neuron layer are universal approximators— i.e., for each continuous function on a bounded domain and for each accuracy $\varepsilon > 0$, we can find a neural network which computes the given function with the desired accuracy. In general, the more accuracy we require, the more neurons we need. So, to achieve perfect accuracy—i.e., exact computations—we will need potentially infinite number of neurons. However, for some "and"- and "or"-operations, we can have perfect accuracy with a limited number of neurons: e.g., the operation $a \cdot b$ can be computed by a 2-neuron network, as

$$a \cdot b = \frac{1}{4} \cdot (a + b)^2 - \frac{1}{4} \cdot (a - b)^2.$$

The operation $a + b - a \cdot b$ can be computed by a 3-neuron network:

$$a + b - a \cdot b = (a + b) - \frac{1}{4} \cdot (a + b)^2 - \frac{1}{4} \cdot (a - b)^2.$$

It would be interesting to describe all such "and"- and "or"-operations. Maybe $a \cdot b$ and $a + b - a \cdot b$ are the only such operations?

References

1. Alvarez, K., Urenda, J.C., Csiszár, O., Csiszár, G., Dombi, J., Eigner, G., Kreinovich, V.: Towards fast and understandable computations: which 'and'- and 'or'-operations can be represented by the fastest (i.e., 1-layer) neural networks? which activations functions allow such representations? Acta Polytech. Hung. **18**(2), 27–45 (2021)
2. Nguyen, H.T., Kreinovich, V., Wojciechowski, P.: Strict Archimedean t-norms and t-conorms as universal approximators. Int. J. Approx. Reason. **18**(3–4), 239–249 (1998)
3. Bishop, C.M.: Pattern Recognition and Machine Learning. Springer, New York (2006)

Chapter 31
Data Processing: Fuzzy Techniques, Part 2

In the previous section, as a side effect of the main result, we also found out which fuzzy operations are the fastest to compute, But what if we use criteria other than computation time?

Empirical studies have shown that in many practical problems, out of all symmetric membership functions, special *distending* functions work best, and out of all hedge operations and negation operations, fractional-linear ones work the best. In this paper, we show that these empirical successes can be explained by natural invariance requirements.

Results from this chapter first appeared in [1].

31.1 Problem: Which Fuzzy Techniques to Use?

Fuzzy techniques: a brief reminder. In many applications, we have knowledge formulated in terms of imprecise ("fuzzy") terms from natural language, like "small", "somewhat small", etc. To translate this knowledge into computer-understandable form, Lotfi Zadeh proposes *fuzzy techniques*; see, e.g., [2–7]. According to these techniques, each imprecise property like "small" can be described by assigning, to each value x of the corresponding quantity, a degree $\mu(x)$ to which, according to the expert, this property is true. These degrees are usually selected from the interval $[0, 1]$, so that 1 corresponds to full confidence, 0 to complete lack of confidence, and values between 0 and 1 describe intermediate degrees of confidence. The resulting function $\mu(x)$ is known as a *membership function*.

In practice, we can only ask finitely many questions to the expert, so we only elicit a few values $\mu(x_1)$, $\mu(x_2)$, etc. Based on these values, we need to estimate the values $\mu(x)$ for all other values x. For this purpose, usually, we select a family of membership functions—e.g., triangular, trapezoidal, etc.—and select a function from this family which best fits the known values.

© The Author(s), under exclusive license to Springer Nature Switzerland AG 2022
J. C. Urenda and V. Kreinovich, *Algebraic Approach to Data Processing*,
Studies in Big Data 115, https://doi.org/10.1007/978-3-031-16780-5_31

For terms like "somewhat small", "very small", the situation is more complicated. We can add different "hedges" like "somewhat", "very", etc., to each property. As a result, we get a large number of possible terms, and it is not realistically possible to ask the expert about each such term. Instead, practitioners estimate the degree to which, e.g., "somewhat small" is true based on the degree to which "small" is true. In other words, with each linguistic hedge, we associate a function h from [0, 1] to [0, 1] that transform the degree to which a property is true into an estimate for the degree to which the hedged property is true.

Similarly to the membership functions, we can elicit a few values $h(x_i)$ of the hedge operation from the experts, and then we extrapolate and/or interpolate to get all the other values of $h(x)$. Usually, a family of hedge operations is pre-selected, and then we select a specific operation from this family which best fits the elicited values $h(x_i)$.

Similarly, instead of asking experts for their degrees of confidence in statements containing negation, such as "not small", we estimate the expert's degree of confidence in these statements based on their degrees of confidence in the positive statements. The corresponding operation $n(x)$ is known as the *negation operation.*

Need to select proper membership functions, proper hedge operations, and proper negation operations. Fuzzy techniques have been successfully applied to many application areas. However, this does not necessarily mean that every time we try to use fuzzy techniques, we get a success story. The success (or not) often depends on which membership functions and which hedge and negation operations we select: for some selections, we get good results (e.g., good control), for other selections, the results are not so good.

What we do in this chapter. There is a lot of empirical data about which selections work better. In this chapter, we provide a general explanation for several of these empirically best selections, an explanation based on the natural concepts of invariance.

Specifically, we explain the following empirically successful selections:

- for symmetric membership functions that describe properties like "small", for which $\mu(x) = \mu(-x)$ and the degree $\mu(|x|)$ decreases with $|x|$, in many practical situations, the most empirically successful are so-called *distending* membership functions, i.e., functions of the type

$$\mu(x) = \frac{1}{1 + a \cdot |x|^b} \qquad (31.1)$$

for some a and b; see, e.g., [8–10];
- among hedge and negation operations, in many practical situations, the most efficient are fractional-linear functions

$$h(x) = \frac{a + b \cdot x}{1 + c \cdot x} \tag{31.2}$$

for some a, b, and c; see, e.g., [11–14].

31.2   Analysis of the Problem

Re-scaling. The variable x describes the value of some physical quantity, such a distance, height, difference in temperatures, etc. When we process these values, we deal with numbers, but numbers depend on the selection of the measuring unit: if we replace the original measuring unit with a new one which is λ times smaller, then all the numerical values will be multiplied by λ: $x \to X = \lambda \cdot x$. For example, 2 m become $2 \cdot 100 = 200$ cm. This transformation from one measuring scale to another is known as *re-scaling*.

Scale-invariance: idea. In many physical situations, the choice of a measuring unit is rather arbitrary. In such situations, all the formulas remain the same no matter what unit we use.

For example, the formula $y = x^2$ for the area of the square with side x remains valid if we replace the unit for measuring sides from meters with centimeters—of course, we then need to appropriately change the unit for y, from square meters to square centimeters. In general, invariance of the formula $y = f(x)$ means that for each re-scaling $x \to X = \lambda \cdot x$, there exists an appropriate re-scaling $y \to Y$ for which the same formula $Y = f(X)$ will be true for the correspondingly re-scaled variables X and Y.

Let us apply this idea to the membership function. It is reasonable to require that the selection of the best membership functions should also not depend on the choice of the unit for measuring the corresponding quantity x. In other words, it is reasonable to require that for each $\lambda > 0$, there should exist some reasonable transformation $y \to Y = T(y)$ of the degree of confidence for which $y = \mu(x)$ implies $Y = \mu(X)$.

So, what are reasonable transformations of the degree of confidence? One way to measure the degree of confidence is to have a poll: ask N experts how many of them believe that a given value x is, e.g., small, count the number M of whose who believe in this, and take the ratio M/N as the desired degree $y = \mu(x)$.

As usual with polls, the more people we ask, the more adequately we describe the general opinion. So, to get a more accurate estimate for $\mu(x)$, it is reasonable to ask more people. When we have a limited number of people to ask, it is reasonable to ask top experts in the field. When we start asking more people, we are thus adding people who are less experienced—and who may therefore be somewhat intimidated by the opinions of the top experts. This intimidation can be expressed in different ways:

- some new people may be too shy to express their own opinion, so they will keep quiet; as a result, if we add A people to the original N, we sill still have the same number M of people voting "yes", and the new ratio will be equal to $Y = \dfrac{M}{N + A}$, i.e., to $Y = a \cdot y$, where $a \overset{\text{def}}{=} \dfrac{N}{N + A}$;
- some new people will be too shy to think on their own and will vote with the majority; so for the case when $M > N/2$, we will have

$$Y = \frac{M + A}{N + A},$$

i.e., since $M = y \cdot N$, we will have

$$Y = \frac{y \cdot N + A}{N + A} = a \cdot y + b,$$

where a is the same as before and $b = \dfrac{A}{N + A}$;
- we may also have a situation in which a certain proportion c of the new people keep quiet while the others vote with the majority; in this case, we have

$$Y = \frac{M + (1 - c) \cdot A}{N + A} = a \cdot y + b,$$

where $a = (1 - c) \cdot \dfrac{A}{N + A}$.

In all these cases, we have a linear transformation $Y = a \cdot y + b$. So, it seems reasonable to identify reasonable transformations with linear ones. We will call the corresponding scale-invariance L-scale-invariance (L for Linear).

What membership functions we consider. We consider symmetric properties, for which $\mu(-x) = \mu(x)$, so it is sufficient to consider only positive values x. Specifically, we properties like "small" for which the degree of confidence decreases with x, going all the way to 0 as x increases. We will call such membership functions s-membership functions (s for small). Thus, we arrive at the following definition.

Definition 31.1 By an s-membership function, we means a function $\mu : (0, \infty) \to [0, 1]$ that, starting with $\mu(0) = 1$, decreases with x (i.e., for which $x_1 > x_2$ implies $\mu(x_1) \geq \mu(x_2)$) and for which $\lim_{x \to \infty} = 0$.

Definition 31.2 We say that an s-membership function $\mu(x)$ is *L-scale-invariant* if for every $\lambda > 0$, there exist values $a(\lambda)$ and $b(\lambda)$ for which $y = \mu(x)$ implies $Y = \mu(X)$, where $X = \lambda \cdot x$ and $Y = a(\lambda) \cdot y + b(\lambda)$.

Unfortunately, this does not solve our problem: as the following result shows, the only L-scale-invariant s-membership functions are constants;

Proposition 31.1 *The only L-scale-invariant s-membership functions are constant functions* $\mu(x) = \text{const}$.

Discussion. What does this result mean? We considered two possible types of reasonable transformations of the degrees of confidence—which both turned out to be linear, and this was not enough. So probably there are other reasonable transformations of degrees of confidence. How can we describe such transformations?

Clearly, if we have a reasonable transformation, then its inverse is also reasonable. Also, a composition of two reasonable transformations should be a reasonable transformation too. So, in mathematical terms, reasonable transformations should form a *group*.

This group should be finite-dimensional, in the sense that different transformations should be uniquely determined by a finite number of parameters—since in the computer, we can store only finitely many parameters. We also know that linear transformations are reasonable. So, we are looking for a finite-dimensional group of transformations from real numbers to real numbers that contains all linear transformations. It is known (see, e.g., [15–17]) that all such transformations are fractional-linear, i.e., have the form

$$\mu \to \frac{a \cdot \mu + b}{1 + c \cdot \mu}.$$

Thus, we arrive at the following definitions.

31.3 Which Symmetric Membership Functions Should We Select: Definitions and the Main Result

Definition 31.3 We say that an s-membership function $\mu(x)$ is *scale-invariant* if for every $\lambda > 0$, there exist values $a(\lambda)$, $b(\lambda)$, and $c(\lambda)$ for which $y = \mu(x)$ implies $Y = \mu(X)$, where $X = \lambda \cdot x$ and

$$Y = \frac{a(\lambda) \cdot y + b(\lambda)}{1 + c(\lambda) \cdot y}.$$

Proposition 31.2 *The only scale-invariant s-membership functions are distending membership functions (31.1).*

Discussion. This result explains the empirical success of distending functions.

31.4   Which Hedge Operations and Negation Operations Should We Select

Discussion. We would like hedging and negation operations $y = h(x)$ to be also invariant, i.e., that for each natural transformation $X = T(x)$, there should be a transformation $Y = S(y)$ for which $y = h(x)$ implies $Y = h(X)$. Now that we know what are natural transformations of membership degrees—they are fractional-linear functions—we can describe this requirement in precise terms.

Definition 31.4 We say that a monotonic function $y = h(x)$ from an open (finite or infinite) interval D to real numbers is *h-scale-invariant* if for every fractional-linear transformation $X = T(x)$, there exists a fractional-linear transformation $Y = S(y)$ for which $y = h(x)$ implies $Y = h(X)$.

Proposition 31.3 *The only h-scale-invariant functions are fractional-linear ones.*

Discussion.

• This result explains the empirical success of fractional-linear hedge operations and negation operations.
• As we show in the proof, it is sufficient to require that a fractional-linear transformation S exists only for all *linear* transformations T.

31.5   Proofs

Proof of Proposition 31.1 We will prove this result by contradiction. Let us assume that the function $\mu(x)$ is not a constant, and let us derive a contradiction.

Substituting the expressions for X, Y, and $y = \mu(x)$ into the formula $Y = \mu(X)$ describing L-scale-invariance, we conclude that for every x and for every λ, we have

$$\mu(\lambda \cdot x) = a(\lambda) \cdot \mu(x) + b(\lambda). \tag{31.3}$$

It is known that monotonic functions are almost everywhere differentiable. Due to the formula (31.3), if a function $\mu(x)$ is differentiable at some point $x = x_0$, it is also differentiable at any point of the type $\lambda \cdot x_0$ for every $\lambda > 0$—and thus, that it is differentiable for all $x > 0$.

Since the function $\mu(x)$ is not constant, there exist values $x_1 \neq x_2$ for which $\mu(x_1) \neq \mu(x_2)$. For these values, the formula (31.3) has the form

$$\mu(\lambda \cdot x_1) = a(\lambda) \cdot \mu(x_1) + b(\lambda); \quad \mu(\lambda \cdot x_2) = a(\lambda) \cdot \mu(x_2) + b(\lambda).$$

Subtracting the two equations, we get

$$\mu(\lambda \cdot x_1) - \mu(\lambda \cdot x_2) = a(\lambda) \cdot (\mu(x_1) - \mu(x_2)),$$

thus

$$a(\lambda) = \frac{\mu(\lambda \cdot x_1) - \mu(\lambda \cdot x_2)}{\mu(x_1) - \mu(x_2)}.$$

Since the function $\mu(x)$ is differentiable, we can conclude that the function $a(\lambda)$ is also differentiable. Thus, the function $b(\lambda) = \mu(\lambda \cdot x) - a(\lambda) \cdot \mu(x)$ is differentiable too.

Since all three functions $\mu(x)$, $a(\lambda)$, and $b(\lambda)$ are differentiable, we can differentiate both sides of the equality (31.3) with respect to λ. If we substitute $\lambda = 1$, we get $x \cdot \mu'(x) = A \cdot \mu(x) + B$, where we denoted $A \overset{\text{def}}{=} a'(1)$, $B \overset{\text{def}}{=} b'(1)$, and $\mu'(x)$, as usual, indicates the derivative. Thus, $x \cdot \dfrac{d\mu}{dx} = A \cdot \mu + B$. We cannot have $A = 0$ and $B = 0$, since then $\mu'(x) = 0$ and $\mu(x)$ would be a constant. Thus, in general, the expression $A \cdot \mu + B$ is not 0, so

$$\frac{d\mu}{A \cdot \mu + B} = \frac{dx}{x}.$$

If $A = 0$, then integration leads to $\dfrac{1}{B} \cdot \mu(x) = \ln(x) + c$, where c_0 is the integration constant. Thus, $\mu(x) = B \cdot \ln(x) + B \cdot c_0$. This expression has negative values for some x, while all the values $\mu(x)$ are in the interval $[0, 1]$. So, this case is impossible.

If $A \neq 0$, then we have $d(A \cdot \mu + B) = A \cdot d\mu$, hence

$$\frac{d(A \cdot \mu + B)}{A \cdot \mu + B} = A \cdot \frac{dx}{x}.$$

Integration leads to $\ln(A \cdot \mu(x) + B) = A \cdot \ln(x) + c_0$. By applying $\exp(z)$ to both sides, we get $A \cdot \mu(x) + B = \exp(c_0) \cdot x^A$, i.e., $\mu(x) = A^{-1} \cdot \exp(c_0) \cdot x^A - B/A$. This expression tends to infinity either for $x \to \infty$ (if $A > 0$) or for $x \to 0$ (if $A < 0$). In both cases, we get a contradiction with our assumption that $\mu(x)$ is always within the interval $[0, 1]$.

The proposition is proven.

Proof of Proposition 31.2 Substituting the expressions for X, Y, and $y = \mu(x)$ into the formula $Y = \mu(X)$ describing scale-invariance, we conclude that for every x and for every λ, we have

$$\mu(\lambda \cdot x) = \frac{a(\lambda) \cdot \mu(x) + b(\lambda)}{1 + c(\lambda) \cdot \mu(x)}. \tag{31.4}$$

Similarly to the Proof of Proposition 31.1, we can conclude that the function $\mu(x)$ is differentiable for all $x > 0$.

Multiplying both sides of the equality (31.4) by the denominator, we conclude that

$$\mu(\lambda \cdot x) + c(\lambda) \cdot \mu(x) \cdot \mu(\lambda \cdot x) = a(\lambda) \cdot \mu(x) + b(\lambda).$$

So, for three different values x_i, we have the following three equations:

$$\mu(\lambda \cdot x_i) + c(\lambda) \cdot \mu(x_i) \cdot \mu(\lambda \cdot x_i) = a(\lambda) \cdot \mu(x_i) + b(\lambda), \quad i = 1, 2, 3.$$

We thus have a system of three linear equations for three unknowns $a(\lambda), b(\lambda)$, and $c(\lambda)$. By Cramer's rule, the solution to such a system is a rational (hence differentiable) function of the coefficients and the right-hand sides. So, since the function $\mu(x)$ is differentiable, we can conclude that the functions $a(\lambda), b(\lambda)$, and $c(\lambda)$ are differentiable as well.

Since all the functions $\mu(x), a(\lambda), b(\lambda)$, and $c(\lambda)$ are differentiable, we can differentiate both sides of the formula (31.4) with respect to λ. If we substitute $\lambda = 1$ and take into account that for $\lambda = 1$, we have $a(1) = 1$ and $b(1) = c(1) = 0$, we get

$$x \cdot \frac{d\mu}{dx} = A \cdot \mu + B - C \cdot \mu^2,$$

where A and B are the same as in the previous proof and $C \overset{\text{def}}{=} c'(1)$.

For $x \to \infty$, we have $\mu(x) \to 0$, so $\mu'(x) \to 0$, and thus $B = 0$ and

$$x \cdot \frac{d\mu}{dx} = A \cdot \mu - C \cdot \mu^2,$$

i.e.,

$$\frac{d\mu}{B \cdot \mu - C \cdot \mu^2} = \frac{dx}{x}. \qquad (31.5)$$

As we have shown in the Proof of Proposition 31.1, we cannot have $C = 0$, so $C \neq 0$. One can easily see that

$$\frac{1}{\mu - \dfrac{B}{C}} - \frac{1}{\mu} = \frac{\dfrac{B}{C}}{\mu \cdot \left(\mu - \dfrac{B}{C} \right)} = \frac{-B}{B \cdot \mu - C \cdot \mu^2}.$$

Thus, by multiplying both sides of equality (31.5) by $-B$, we get

$$\frac{d\mu}{\mu - \dfrac{B}{C}} - \frac{d\mu}{\mu} = -B \cdot \frac{dx}{x}.$$

Integrating both sides, we get

$$\ln\left(\mu(x) - \frac{B}{C}\right) - \ln(\mu) = -B \cdot \ln(x) + c_0.$$

By applying $\exp(z)$ to both sides, we get

$$\frac{\mu(x) - \dfrac{B}{C}}{\mu(x)} = C_0 \cdot x^{-B}$$

for some constant C_0, i.e.,

$$1 - \frac{B/C}{\mu} = C_0 \cdot x^{-B},$$

hence

$$\frac{B/C}{\mu} = 1 - C_0 \cdot x^{-B}$$

and

$$\mu(x) = \frac{B/C}{1 - C_0 \cdot x^{-B}}.$$

From the condition that $\mu(0) = 1$, we conclude that $B < 0$ and $B/C = 1$. From the condition that $\mu(x) \leq 1$, we conclude that $C_0 < 0$. Thus, we get the desired formula

$$\mu(x) = \frac{1}{1 + |C_0| \cdot x^{|B|}}.$$

The proposition is proven.

Proof of Proposition 31.3 For constant functions the statement is trivial, since every constant function is fractional-linear. Therefore, it is sufficient to prove for non-constant functions $h(x)$.

Similarly to the Proof of Proposition 31.2, we can prove that the function $h(x)$ is differentiable. Let $x \in D$, and let λ and x_0 from an open neighborhood of 1 and 0 respectively be such that $\lambda \cdot x \in D$ and $x + x_0 \in D$. Since the function $h(x)$ is h-scale-invariant, there exist fractional-linear transformations for which

$$h(x + x_0) = \frac{a(x_0) \cdot h(x) + b(x_0)}{1 + c(x_0) \cdot h(x)} \tag{31.6}$$

and

$$h(\lambda \cdot x) = \frac{d(\lambda) \cdot h(x) + e(\lambda)}{1 + f(\lambda) \cdot h(x)}. \tag{31.7}$$

Similarly to the Proof of Proposition 31.2, we can prove that the functions $a(x_0)$, ..., are differentiable. Similar to the Proof of Proposition 31.2, we can differentiate the formula (31.7) with respect to λ and take $\lambda = 1$, then we get:

$$x \cdot h' = D \cdot h + E - F \cdot h^2. \tag{31.8}$$

Similarly, differentiating the formula (31.6) with respect to x_0 and taking $x_0 = 0$, we get:

$$h' = A \cdot h + B - C \cdot h^2. \tag{31.9}$$

Let us consider two cases: $C \neq 0$ and $C = 0$.

Let us first consider the case when $C \neq 0$. By completing the square, we get $h' = A \cdot h + B - C \cdot h^2 = \widehat{A} - C \cdot (h - h_0)^2$ for some $\widehat{A}$ and h_0, i.e.,

$$h' = \widehat{A} - C \cdot H^2, \tag{31.10}$$

where $H \stackrel{\text{def}}{=} h - h_0$. Substituting $h = H + h_0$ into the right-hand of the formula (31.8), we conclude that

$$x \cdot h' = \widehat{D} \cdot H + \widehat{E} - F \cdot H^2 \tag{31.11}$$

for some constants $\widehat{D}$ and $\widehat{E}$. Dividing (31.11) by (31.10), we get

$$x = \frac{\widehat{D} \cdot H + \widehat{E} - F \cdot H^2}{\widehat{A} - C \cdot H^2}, \tag{31.12}$$

so

$$\frac{dx}{dH} = \frac{(\widehat{D} - 2F \cdot H) \cdot (\widehat{A} - C \cdot H^2) - (\widehat{D} \cdot H + \widehat{E} - F \cdot H^2) \cdot (-2C \cdot H)}{(\widehat{A} - C \cdot H^2)^2} =$$

$$\frac{\widehat{A} \cdot \widehat{D} - 2(\widehat{A} \cdot F - C \cdot \widehat{E}) \cdot H + C \cdot \widehat{D} \cdot H^2}{(\widehat{A} - C \cdot H^2)^2}. \tag{31.13}$$

On the other hand,

$$\frac{dx}{dH} = \frac{1}{\dfrac{dH}{dx}} = \frac{1}{\widehat{A} - C \cdot H^2}. \tag{31.14}$$

The right-hand sides of the formulas (31.13) and (31.14) must be equal, so for all H, we have

$$\widehat{A} \cdot \widehat{D} - 2(\widehat{A} \cdot F - C \cdot \widehat{E}) \cdot H + C \cdot \widehat{D} \cdot H^2 = \widehat{A} - C \cdot H^2.$$

Since the two polynomials of H are equal, the coefficients at 1, H, and H^2 must coincide.

Comparing the coefficients at H^2, we get $C \cdot \widehat{D} = -C$. Since $C \neq 0$, we conclude that $\widehat{D} = -1$. Comparing the coefficients at 1, we get $\widehat{A} \cdot \widehat{D} = \widehat{A}$, i.e., $-\widehat{A} = \widehat{A}$ and thus $\widehat{A} = 0$. Comparing the coefficients at H and taking into account that $\widehat{A} = 0$, we get $0 = \widehat{A} \cdot F - C \cdot \widehat{E} = -C \cdot \widehat{E}$. Since $C \neq 0$, this implies $\widehat{E} = 0$. So, the formula (31.12) takes the form

$$x = \frac{\widehat{D} \cdot H - F \cdot H^2}{-C \cdot H^2} = \frac{\widehat{D} - F \cdot H}{-C \cdot H}.$$

Thus x is a fractional-linear function of H, hence H (and therefore $h = H + h_0$) is also a fractional-linear function of x.

Let us now consider the case when $C = 0$. In this case, $h' = A \cdot h + B$ and $x \cdot h' = D \cdot h + E - F \cdot h^2$, thus

$$x = \frac{x \cdot h'}{h'} = \frac{D \cdot h + E - F \cdot h^2}{A \cdot h + B}.$$

If $F = 0$, then x is a fractional-linear function of $h(x)$ and hence, h is also a fractional-linear function of x.

So, it is sufficient to consider the case when $F \neq 0$. In this case, by completing the square, we can find constants $\widetilde{D}$, h_0, and $\widehat{B}$ for which, for $H = h - h_0$, we have

$$x \cdot h' = D \cdot h + E - F \cdot h^2 = \widetilde{D} - F \cdot H^2 \tag{31.15}$$

and

$$h' = A \cdot h + B = A \cdot H + \widehat{B}. \tag{31.16}$$

Dividing (31.15) by (31.16), we have

$$x = \frac{\widetilde{D} - F \cdot H^2}{A \cdot H + \widehat{B}}. \tag{31.17}$$

Thus,

$$\frac{dx}{dH} = \frac{(-2F \cdot H) \cdot (A \cdot H + \widehat{B}) - (\widetilde{D} - F \cdot H^2) \cdot A}{(A \cdot H + \widehat{B})^2}$$

$$= \frac{-A \cdot \widetilde{D} - 2\widehat{B} \cdot F \cdot H - A \cdot F \cdot H^2}{(A \cdot H + \widehat{B})^2}.$$

On the other hand,

$$\frac{dx}{dH} = \frac{1}{\dfrac{dH}{dx}} = \frac{1}{A \cdot H + \widehat{B}}.$$

By equating the two expressions for the derivative and multiplying both sides by $(A \cdot H + \widehat{B})^2$, we conclude that

$$-A \cdot \widehat{D} - 2\widehat{B} \cdot F \cdot H - A \cdot F \cdot H^2 = A \cdot H + \widehat{B},$$

thus $A \cdot F = 0$, $A = -2\widehat{B} \cdot F$, and $-A \cdot \widehat{D} = \widehat{B}$. If $A = 0$, then we have $\widehat{B} = 0$, so $h' = 0$ and h is a constant—but we consider the case when the function $h(x)$ is not a constant. Thus, $A \neq 0$, hence $F = 0$, and the formula (31.17) describes x as a fractional-linear function of H.

In both cases $C \neq 0$ and $C = 0$, we obtain an expression of x in terms of H (hence h) that is fractional-linear. Since the inverse of a fractional linear is fractional-linear, the function $h(x)$ is also fractional-linear.

The proposition is proven.

References

1. Urenda, J.C., Csiszár, O., Csiszár, G., Dombi, J., Eigner, G., Kreinovich, V.: Natural invariance explains empirical success of specific membership functions, hedge operations, and negation operations. In: Proceedings of the Annual Conference of the North American Fuzzy Information Processing Society NAFIPS'2020, Redmond, Washington (2020). Accessed from 20–22 Aug 2020
2. Belohlavek, R., Dauben, J.W., Klir, G.J.: Fuzzy Logic and Mathematics: A Historical Perspective. Oxford University Press, New York (2017)
3. Klir, G., Yuan, B.: Fuzzy Sets and Fuzzy Logic. Prentice Hall, Upper Saddle River, New Jersey (1995)
4. Mendel, J.M.: Uncertain Rule-Based Fuzzy Systems: Introduction and New Directions. Springer, Cham, Switzerland (2017)
5. Nguyen, H.T., Walker, C.L., Walker, E.A.: A First Course in Fuzzy Logic. Chapman and Hall/CRC, Boca Raton, Florida (2019)
6. Novák, V., Perfilieva, I., Močkoř, J.: Mathematical Principles of Fuzzy Logic. Kluwer, Boston, Dordrecht (1999)
7. Zadeh, L.A.: Fuzzy sets. Inf. Control **8**, 338–353 (1965)
8. Dombi, J., Hussain, A.: Interval type-2 fuzzy control using the distending function. In: Tallón-Ballesteros, A.J. (ed.) Proceedings of the 5th International Conference on Fuzzy Systems, Kitakyushu, Japan, pp. 705–714 (2019). Accessed from 18–21 Oct 2019
9. Dombi, J., Hussain, A.: Data-driven arithmetic fuzzy control using the distending function. In: Proceedings of the International Conference on Human Interaction and Emerging Technologies IHIET'2019, Nice, France, pp. 215–221 (2019). Accessed from 22–24 Aug 2019
10. Dombi, J., Hussain, A.: A new approach to fuzzy control using the distending function. J. Process Control **86**, 16–29 (2020)
11. Csiszar, O., Dombi, J.: Generator-based modifiers and membership functions in nilpotent operator systems. In: Proceedings of the IEEE International Work Conference on Bioinspired Intelligence IWOBI'2019, Budapest, Hungary, pp. 99–105 (2019). Accessed from 3–5 July 2019
12. Dombi, J., Csiszar, O.: Implications in bounded systems. Inf. Sci. **283**, 229–240 (2014)
13. Dombi, J., Csiszar, O.: The general nilpotent operator system. Fuzzy Sets Syst. **261**, 1–19 (2015)
14. Dombi, J., Csiszar, O.: Equivalence operators in nilpotent systems. Fuzzy Sets Syst. **299**, 113–129 (2016)

15. Guillemin, V.M., Sternberg, S.: An algebraic model of transitive differential geometry. Bull. Am. Math. Soc. **70**(1), 16–47 (1964)
16. Nguyen, H.T., Kreinovich, V.: Applications of Continuous Mathematics to Computer Science. Kluwer, Dordrecht (1997)
17. Singer, I.M., Sternberg, S.: Infinite groupsof Lie and Cartan, Part I. J. d'Analyse Mathematique **15**, 1–113 (1965)

Chapter 32
Data Processing: Fuzzy Techniques, Part 3

In this chapter and in the following chapter, we deal with two auxiliary fuzzy-related issues.

In this chapter, we deal with the fact that while in the traditional fuzzy logic, experts' degrees of confidence are described by numbers from the interval [0, 1], clearly, not all the numbers from this interval are needed. Indeed, in the whole history of the Universe, there will be only countably many statements and thus, only countably many possible degree, while the interval [0, 1] is uncountable. It is therefore interesting to analyze what is the set S of actually used values. The answer depends on the choice of "and"-operations (t-norms) and "or"-operations (t-conorms). For the simplest pair of min and max, any finite set will do—as long as it is closed under negation $1 - a$. For the next simplest pair—of algebraic product and algebraic sum—we prove that for a finitely generated set, if the "and"-operation is exact, then the "or"-operation is almost always approximate, and vice versa. For other "and"- and "or"-operations, the situation can be more complex.

Results from this chapter first appeared in [1].

32.1 Problem: Which Fuzzy Degrees to Use?

Need for fuzzy degrees: a brief reminder. Computers are an important part of our lives. They help us understand the world, they help us make good decisions. It is desirable to make sure that these computers possess as much of our own knowledge as possible.

Some of this knowledge is precise; such a knowledge is relatively easy to describe in computer-understandable terms. However, a significant part of our knowledge is described by using imprecise ("fuzzy") words from natural language. For example, to design better self-driving cars, it sounds reasonable to ask experienced drivers

J. C. Urenda and V. Kreinovich, *Algebraic Approach to Data Processing*, Studies in Big Data 115, https://doi.org/10.1007/978-3-031-16780-5_32
185

what they do in different situations, and implement the corresponding rules in the car's computer. The problem with this idea is that expert drivers usually describe their rules by saying something like "If the car in front of you is close, and it slows down a little bit, then ...". Here, "close", "a little bit", etc. are imprecise words.

To translate such knowledge into computer-understandable terms, Lotfi Zadeh invented *fuzzy logic*, in which each imprecise terms like "close" is described by assigning, to each possible value x of the corresponding quantity (in this case, distance), a degree to which x satisfies the property under consideration (in this case, to what exact, x is close); see, e.g., [2–7]. In the original formulation of fuzzy logic, the degrees are described by numbers from the interval [0, 1], so that:

- 1 means that we are absolutely sure about the corresponding statement,
- 0 means that we are sure that this statement is false, and
- intermediate degrees corresponding to intermediate degrees of certainty.

Need for operations on fuzzy degrees, i.e., for fuzzy logic: a brief reminder. Our rules often use logical connectives like "and" and "or". In the above example of the car-related statement, the person used "and" and "if-then" (implication). Other statements use negation or "or".

In the ideal world, we should ask each expert to describe his or her degree of confidence in each such statement—e.g., in the statement that "the car in front of you is close, and it slows down a little bit". However, here is a problem: to describe each property like "close" for distance or "a little bit" for a change in speed, it is enough to list possible values of one variable. For a statement about two variables— as above—we already need to consider all possible pairs of values. So, if we consider N possible values of each variable, we need to ask the expert N^2 questions. If our statement involves three properties—which often happens—we need to consider N^3 possible combinations, etc. With a reasonably large N, this quickly becomes impossible to ask the expert all these thousands (and even millions) of questions. So, instead of explicitly asking all these questions about composite statements like $A \& B$, we need to be able to estimate the expert's degree of confidence in such a statement based on his or her known degrees of confidence a and b in the original statements A and B.

A procedure that transform these degrees a and b into the desired estimate for the degree of confidence in $A \& B$ is known as an *"and"-operation* (or, for historical reasons, a *t-norm*). We will denote this procedure by $f_\&(a, b)$. Similarly, a procedure corresponding to "or" is called an *"or"-operation* or a *t-conorm*; it will be denoted by $f_\vee(a, b)$. We can also have negation operations $f_\neg(a)$, implication operation $f_\rightarrow(a, b)$, etc.

Simple examples of operations on fuzzy degrees. In his very first paper on fuzzy logic, Zadeh considered the two simplest possible "and"-operations $\min(a, b)$ and $a \cdot b$, the simplest negation operation $f_\neg(a) = 1 - a$, and the simplest possible "or"- operations $\max(a, b)$ and $a + b - a \cdot b$.

It is easy to see that the corresponding "and"- and "or"-operations form two dual pairs, i.e., pairs for which $f_\vee(a, b) = f_\neg(f_\&(f_\neg(a), f_\neg(b)))$—this reflects the fact that in our reasoning, $a \vee b$ is indeed usually equivalent to $\neg(\neg a \,\&\, \neg b)$. Indeed, for example, what does it mean that a dish contains either pork or alcohol (or both)? It simply means that it is not true that this is an alcohol-free and pork-free dish.

Both pairs of operations can be derived from the requirement of the smallest sensitivity to changes in a and b (see, e.g., [5, 8, 9])–which makes sense, since experts can only mark their degree of confidence with some accuracy, and we do not want the result of, e.g., "and"-operation drastically change if we replace the original degree 0.5 with a practically indistinguishable degree 0.51. If we require that the worst-case change in the result of the operation be as small as possible, we get $\min(a, b)$ and $\max(a, b)$. If we require that the mean squares value of the change be as small as possible, we get $a \cdot b$ and $a + b - a \cdot b$.

Implication $A \rightarrow B$ means that, if we know A and we know that implication is true, then we can conclude B. In other words, implication is the weakest of all statements C for which $A \,\&\, C$ implies B. So, if we know the degrees of confidence a and b in the statements A and B, then a reasonable definition of the implication $f_\vee(a, b)$ is the smallest degree c for which $f_\&(a, c) \geq b$. In this sense, implication is, in some reasonable sense, an inverse to the "and"-operation. In particular, when the "and"-operation is multiplication $f_\&(a, b) = a \cdot b$, the implication operation is simply division: $f_\rightarrow(a, b) = b/a$ (if $b \leq a$).

Not all values from the interval $[0, 1]$ **make sense.** While it is reasonable to use numbers from the interval $[0, 1]$ to describe the corresponding degrees, the inverse is not true—not every number from the interval $[0, 1]$ makes sense as a degree. Indeed, whatever degree we use corresponds to some person's informal description of his or her degree of confidence. Whatever language we use, there are only countably many words, while, as is well known, the set of all real numbers from an interval is uncountable.

Usually, we have a finite set of basic degrees, and everything else is obtained by applying some logical operations. A natural question is: what can we say about the resulting—countable—sets of actually used values? This is a general question to which, in this chapter, we provide a partial answer.

Simplest case: min **and** max. The simplest case if when have $f_\&(a, b) = \min(a, b)$, $f_\neg(a) = 1 - a$, and $f_\vee(a, b) = \max(a, b)$. In this case, if we start with a finite set of degrees $a_1, \ldots, a_n$, then we add their negations $1 - a_1, \ldots, 1 - a_n$, and that, in effect, is it: $\min(a, b)$ and $\max(a, b)$ do not generate any new values, they just select one of the two given ones (a or b).

What about $a \cdot b$ **and** $a + b - a \cdot b$. What about the next simplest pair of operations? Since the product is the simplest of the two, let us start with the product. Again, we start with a finite set of degrees $a_1, \ldots, a_n$. We can also consider their negations $a_{n+1} \stackrel{\text{def}}{=} 1 - a_1, \ldots, a_{n+i} \stackrel{\text{def}}{=} 1 - a_i, \ldots, a_{2n} \stackrel{\text{def}}{=} 1 - a_n$.

If we apply "and"-operation to these values, we get products, i.e., values the type

$$a_1^{k_1} \cdot \ldots \cdot a_{2n}^{k_{2n}} \tag{32.1}$$

for integers $k_i \geq 0$. If we also allow implication—i.e., in this case, division—then we get values of the same type (32.1), but with integers k_i being possibly negative.

The set of all such values is generated based on the original finite set of values. Thus, we can say that this set is *finitely generated*.

Every real number can be approximated, with any given accuracy, by a rational number. Thus, without losing generality, we can assume that all the values a_i are rational numbers—i.e., ratios of two integers.

Since for dual operations, the result of applying the "or"-operation is the negation of the result of applying the "and"-operation—to negations of a and b—a natural question is: if we take values of type (32.1), how many of their negations are also of the same type? This is a question that we study in this chapter.

32.2   Definitions and the Main Result

Definition 32.1 By a *finitely generated set* of fuzzy degrees, we mean a set S of values of the type (32.1) from the interval $[0, 1]$, where $a_1, \ldots, a_n$ are given rational numbers, $a_{n+i} = 1 - a_i$, and $k_1, \ldots, k_{2n}$ are arbitrary integers.

Examples. If we take $n = 1$ and $a_1 = 1/2$, then $a_2 = 1 - a_1 = 1/2$, so all the values of type (32.1) are 1/2, 1/4, 1/8, etc. Here, only for one number $a_1 = 1/2$, the negation $1 - a_1$ belongs to the same set.

If we take $a_1 = 1/3$ and $a_2 = 2/3$, then we have more than one number s from the set S for which its negation $1 - s$ is also in S:

- we have $1/4 = a_1^2 \cdot a_2^{-2} \in S$ for which $1 - 1/4 = 3/4 = a_1 \cdot a_2^{-2} \in S$; and
- we have $1/9 = a_1^2 \in S$ and $1 - 1/9 = 8/9 = a_1^{-1} \cdot a_2^3 \in S$.

Proposition 32.1 *For each finitely generated set S of fuzzy degrees, there are only finitely many element $s \in S$ for which $1 - s \in S$.*

Proof Is, in effect, contained in [10–18], where the values $s \in S$ are called *S-units* and the desired formula $s + s' = 1$ for $s, s' \in S$ is known as the *S-unit equation*.

Historical comment. The history of this mathematical result is unusual (see, e.g., [19]): the corresponding problem was first analyzed by Axel Thue in 1909, it was implicitly proven by Carl Lugwig Siegel in 1929, then another implicit proof was made by Kurt Mahler in 1933—but only reasonably recently this result was explicitly formulated and explicitly proven.

Discussion. Proposition 32.1 says that for all but finitely many ("almost all") values $s \in S$, the negation $1 - s$ is outside the finitely generated set S.

Since, as we have mentioned, to get an "or"-operation out of "and" requires negation, this means that while for this set, "and"-operation is exact, the corresponding "or"-operation almost always leads us to a value outside S. So, if we restrict ourselves to the finitely generated set S, we can only represent the results of "or"-operation approximately.

In other words, if "and" is exact, then "or" is almost always approximate. Due to duality between "and"- and "or", we can also conclude that if "or" is exact, then "and" is almost always approximate.

Computational aspects. The formulation of our main result sounds like (too) abstract mathematics: there exists finitely many such values s; but how can we find them? Interesting, there exists a reasonably efficient algorithms for finding such values; see, e.g., [20].

Relation to probabilities. Our current interest is in fuzzy logic, but it should be mentioned that a similar results holds for the case of probabilistic uncertainty, when, instead of degrees of confidence, we consider possible probability values a_i. In this case:

- if an event has probability a, then its negation has probability $1 - a$;
- if two independent events have probabilities a and b, then the probably that both events will happen is $a \cdot b$; and
- if an event B is a particular case of an event A, then the conditional probability $P(B \mid A)$ is equal to b/a.

Thus, in the case of probabilistic uncertainty, it also makes sense to consider multiplication and division operations—and thus, to consider sets which are closed under these operations.

32.3 How General Is This Result?

Formulation of the problem. In the previous section, we considered the case when $f_\&(a, b) = a \cdot b$ and $f_\neg(a) = 1 - a$. What if we consider another pair of operations, will the result still be true?

For example, is it true for *strict Archimedean* "and"-operations?

Analysis of the problem. It is known that every strict Archimedean "and"-operation is equivalent to $f_\&(a, b) = a \cdot b$—namely, we can reduce it to the product by applying an appropriate strictly increasing re-scaling $r : [0, 1] \to [0, 1]$; see, e.g., [3, 5].

Thus, without losing generality, we can assume that the "and"-operation is exactly the product $f_\&(a, b) = a \cdot b$, but the negation operation may be different—as long as $f_\neg(f_\neg(a)) = a$ for all a.

Result of this section. It turns out that there are some negation operations for which the above result does not hold.

Proposition 32.2 *For each finitely generated set S—with the only exception of the set generated by a single value 1/2—there exists a negation operation $f_\neg(a)$ for which, for infinitely many $s \in S$, we have $f_\neg(s) \in S$.*

Proof When at least one of the original values a_i is different from 1/2, this means that the fractions a_i and $1 - a_i$ have different combinations of prime numbers in their numerators and denominators. In this case, for every $\varepsilon > 0$, there exists a number $s \in S$ for which $1 - \varepsilon < s < 1$.

We know that one of the original values a_i is different from 1/2. Without losing generality, let us assume that this value is 1/2. If $a_1 > 1/2$, then $1 - a_1 < 1/2$. So, again without losing any generality, we can assume that $a_1 < 1/2$.

Let us now define two monotonic sequences p_n and q_n. For the first sequence, we take the values

$$ p_0 = 1/2 > p_1 = a_1 > p_2 = a_1^2 > p_3 = a_1^3 > \cdots $$

The second sequence is defined iteratively:

- As q_0, we take $q_0 = 1/2$.
- As q_1, let us select some number (smaller than 1) from the set S which is greater than or equal to $1 - p_1$.
- Once the values $q_1, \ldots, q_k$ have been selected, we select, as q_{k+1}, a number (smaller than 1) from the set S which is larger than q_k and larger than $1 - p_{k+1}$, etc.

For values $s \le 0.5$, we can then define the negation operation as follows:

- for each k, we have $f_\neg(p_k) = q_k$ and
- it is linear for $p_{k+1} < s < p_k$, i.e.

$$ f_\neg(s) = q_{k+1} + (s - p_{k+1}) \cdot \frac{q_{k+1} - q_k}{p_{k+1} - p - k}. $$

The resulting function maps the interval $[0, 0.5]$ to the interval $[0.5, 1]$. For values $s \ge 0.5$, we can define $f_\neg(a)$ as the inverse function to this.

32.4   What If We Allow Unlimited Number of "And"-Operations and Negations: Case Study

Formulation of the problem. In the previous sections, we allowed an unlimited application of "and"-operation and implication. What if instead, we allow an unlimited application of "and"-operation and negation?

Here is our related result.

Proposition 32.3 *The set S of degrees that can be obtained from 0, 1/2, and 1 by using "and"-operation $f_\&(a, b) = a \cdot b$ and negation $f_\neg(a) = 1 - a$ is the set of all binary-rational numbers, i.e., all numbers of the type $p/2^k$ for natural numbers p and k for which $p \le 2^k$.*

Proof Clearly, the product of two binary-rational numbers is binary-rational, and 1 minus a binary-rational number is also a binary-rational number. So, all elements of the set S are binary-rational.

To complete the proof, we need to show that every binary-rational number $p/2^k$ belongs to the set S, i.e., can be obtained from 1/2 by using multiplication and $1 - a$. We will prove this result by induction over k.

For $k = 1$, this means that 0, 1/2, and 1 belong to the set S—and this is clearly true, since S consists of all numbers that can be obtained from these three, these three numbers included.

Let us assume that this property is proved for k. Then, for $p \leq 2^k$, each element $p/2^{k+1}$ is equal to the product $(1/2) \cdot (p/2^k)$ of two numbers from the set S and thus, also belongs to S. For $p > 2^k$, we have

$$p/2^{k+1} = 1 - (2^{k+1} - p)/2^{k+1}.$$

Since $p > 2^k$, we have $2^{k+1} - p < 2^k$ and thus, as we have just proved, $(2^{k+1} - p)/2^{k+1} \in S$. So, the ratio $p/2^{k+1}$ is obtained by applying the negation operation to a number from the set S and is, therefore, itself an element of the set S.

The induction step is proven, and so is the proposition.

Comment. If we also allow implication $f_{\&}(a, b) = b/a$, then we will get all possible rational numbers p/q from the interval $[0, 1]$. Indeed, if we pick k for which $q < 2^k$, then for $a = q/2^k$ and $b = p/2^k$, we get $b/a = p/q$.

References

1. Urenda, J., Kosheleva, O., Kreinovich, V.: Finitely generated sets of fuzzy values: if 'and' is exact, then 'or' is almost always approximate and vice versa. In: Ceberio, M., Kreinovich, V. (eds.) How Uncertainty-Related Ideas Can Provide Theoretical Explanation for Empirical Dependencies, pp. 133–140. Springer, Cham (2021)
2. Belohlavek, R., Dauben, J.W., Klir, G.J.: Fuzzy Logic and Mathematics: A Historical Perspective. Oxford University Press, New York (2017)
3. Klir, G., Yuan, B.: Fuzzy Sets and Fuzzy Logic. Prentice Hall, Upper Saddle River (1995)
4. Mendel, J.M.: Uncertain Rule-Based Fuzzy Systems: Introduction and New Directions. Springer, Cham (2017)
5. Nguyen, H.T., Walker, C.L., Walker, E.A.: A First Course in Fuzzy Logic. Chapman and Hall/CRC, Boca Raton (2019)
6. Novák, V., Perfilieva, I., Močkoř, J.: Mathematical Principles of Fuzzy Logic. Kluwer, Boston (1999)
7. Zadeh, L.A.: Fuzzy sets. Information and Control **8**, 338–353 (1965)
8. Nguyen, H.T., Kreinovich, V., Tolbert, D.: On robustness of fuzzy logics. In: Proceedings of the 1993 IEEE International Conference on Fuzzy Systems FUZZ-IEEE'93, San Francisco, California, vol. 1, pp. 543–547 (1993)
9. Tolbert, D.: Finding "and" and "or" operations that are least sensitive to change in intelligent control. Master's Thesis, Department of Computer Science, University of Texas at El Paso (1994)

10. Baker, A., Wüstholz, G.: Logarithmic Forms and Diophantine Geometry. Cambridge University Press, Cambridge (2007)
11. Bombieri, E., Gubler, W.: Heights in Diophantine Geometry. Cambridge University Press, Cambridge (2006)
12. Everest, G., van der Poorten, A., Shparlinski, I., Ward, Th.: Recurrence Sequences. American Mathematical Society, Providence (2003)
13. Lang, S.: Elliptic Curves: Diophantine Analysis. Springer, Berlin (1978)
14. Lang, S.: Algebraic Number Theory. Springer, Berlin (1986)
15. Malmskog, B., Rasmussen, C.: Picard curves over $\mathbb{Q}$ with good reduction away from 3. Lond. Math. Soc. (LMS) J. Comput. Math. **19**(2), 382–408 (2016)
16. Neukirch, J.: Class Field Theory. Springer, Berlin (1986)
17. Smart, N.P.: The solution of triangularly connected decomposable form equations. Math. Comput. **64**(210), 819–840 (1995)
18. Smart, N.P.: The Algorithmic Resolution of Diophantine Equations. London Mathematical Society, London (1998)
19. Mackenzie, D.: Needles in an infinite haystack. In: Mackenzie, D. (ed.) What's Happening in the Mathematical Sciences, vol. 11, pp. 123–136. American Mathematical Society, Providence (2019)
20. Alvarado, A., Koutsianas, A., Malmskog, B., Rasmussen, C., Roe, D., Vincent, C., West, M.: Solve S-unit equation x + y = 1 - Sage Reference Manual v8.7: Algebraic Numbers and Number Fields (2019). http://doc.sagemath.org/html/en/reference/number_fields/sage/rings/number_field/S_unit_solver.html

Chapter 33
Data Processing: Fuzzy Techniques, Part 4

In this chapter, we deal with yet another auxiliary fuzzy-related issue.

Specifically, we deal with fact that while in the usual 2-valued logic, from the purely mathematical viewpoint, there are many possible binary operations, in commonsense reasoning, we only use a few of them. Why? In this chapter, we show that fuzzy logic can explain the usual choice of logical operations in 2-valued logic.

Results from this chapter first appeared in [1].

33.1 Problem: How to Explain Commonsense Reasoning

In 2-valued logic, there are many possible logical operations: reminder. In the usual 2-valued logic, in which each variable can have two possible truth values— 0 (false) or 1 (true), for each n, there are many possible n logical operations, i.e., functions $f : \{0, 1\}^n \to \{0, 1\}$. To describe each such function, we need to describe, for each of 2^n boolean vectors $(a_1, \ldots, a_n)$, whether the resulting value $f(a_1, \ldots, a_n)$ is 0 or 1.

Case of unary operations. For unary operations, i.e., operations corresponding to $n = 1$, we need to describe two values: $f(0)$ and $f(1)$. For each of these two values, there are 2 possible options, so overall, we have $2 \cdot 2 = 2^2 = 4$ possible unary operations:

- the case when $f(0) = f(1) = 0$ corresponds to a constant $f(a) \equiv 0$;
- the case when $f(0) = 0$ and $f(1) = 1$ corresponds to the identity function

$$f(a) = a;$$

- the case when $f(0) = 1$ and $f(1) = 0$ corresponds to negation $f(a) = \neg a$; and

© The Author(s), under exclusive license to Springer Nature Switzerland AG 2022
J. C. Urenda and V. Kreinovich, *Algebraic Approach to Data Processing*,
Studies in Big Data 115, https://doi.org/10.1007/978-3-031-16780-5_33

- the case when $f(0) = f(1) = 1$ corresponds to a constant function

$$f(a) = 1.$$

The only non-trivial case is negation, and it is indeed actively used in our logical reasoning.

Case of binary operations. For binary operations, i.e., operations corresponding to $n = 2$, we need to describe four values $f(0, 0)$, $f(0, 1)$, $f(1, 0)$, and $f(1, 1)$. For each of these four values, there are 2 possible options, so overall, we have $2^4 = 16$ possible binary operations:

- the case when $f(0, 0) = f(0, 1) = f(1, 0) = f(1, 1) = 0$ corresponds to a constant function $f(a, b) \equiv 0$;
- the case when $f(0, 0) = f(0, 1) = f(1, 0) = 0$ and $f(1, 1) = 1$ corresponds to "and" $f(a, b) = a \& b$;
- the case when $f(0, 0) = f(0, 1) = 0$, $f(1, 0) = 1$, and $f(1, 1) = 0$, corresponds to $f(a, b) = a \& \neg b$;
- the case when $f(0, 0) = f(0, 1) = 0$ and $f(1, 0) = f(1, 1) = 1$, corresponds to $f(a, b) = a$;
- the case when $f(0, 0) = 0$, $f(0, 1) = 1$, and $f(1, 0) = f(1, 1) = 0$, corresponds to $f(a, b) = \neg a \& b$;
- the case when $f(0, 0) = 0$, $f(0, 1) = 1$, $f(1, 0) = 0$, and $f(1, 1) = 1$, corresponds to $f(a, b) = b$;
- the case when $f(0, 0) = 0$, $f(0, 1) = 1$, $f(1, 0) = 1$, and $f(1, 1) = 0$, corresponds to exclusive "or" (= addition modulo 2) $f(a, b) = a \oplus b$;
- the case when $f(0, 0) = 0$ and $f(0, 1) = f(1, 0) = f(1, 1) = 1$, corresponds to "or" $f(a, b) = a \vee b$;
- the case when $f(0, 0) = 1$ and $f(0, 1) = f(1, 0) = f(1, 1) = 0$, corresponds to $f(a, b) = \neg a \& \neg b$;
- the case when $f(0, 0) = 1$, $f(0, 1) = f(1, 0) = 0$, and $f(1, 1) = 1$, corresponds to equivalence (equality) $f(a, b) = a \equiv b$;
- the case when $f(0, 0) = 1$, $f(0, 1) = 0$, $f(1, 0) = 1$, and $f(1, 1) = 0$, corresponds to $f(a, b) = \neg b$;
- the case when $f(0, 0) = 1$, $f(0, 1) = 0$, and $f(1, 0) = f(1, 1) = 1$, corresponds to $f(a, b) = a \vee \neg b$ or, equivalently, to the implication

$$f(a, b) = b \to a;$$

- the case when $f(0, 0) = f(0, 1) = 1$, and $f(1, 0) = f(1, 1) = 0$, corresponds to $f(a, b) = \neg a$;
- the case when $f(0, 0) = f(0, 1) = 1$, $f(1, 0) = 0$, and $f(1, 1) = 1$, corresponds to $f(a, b) = \neg a \vee b$, or, equivalently, to the implication

$$f(a, b) = a \to b;$$

- the case when $f(0, 0) = f(0, 1) = f(1, 0) = 1$, and $f(1, 1) = 0$, corresponds to $f(a, b) = \neg a \vee \neg b$;
- the case when $f(0, 0) = f(0, 1) = f(1, 0) = f(1, 1) = 1$, corresponds to a constant function $f(a, b) \equiv 1$.

In commonsense reasoning, we only some of the binary operations. Out of the above 16 operations,

- two are constants: $f(a, b) = 0$ and $f(a, b) = 1$, and
- four are actually unary: $f(a, b) = a$, $f(a, b) \neg a$, $f(a, b) = b$, and

$$f(a, b) = \neg b.$$

In addition to these $2 + 4 = 6$ operations, there are also 10 non-constant and non-unary binary logical operations:

- six named operations "and", "or", exclusive "or", equivalence, and two implications ($a \to b$ and $b \to a$), and
- four usually un-named logical operations

$$\neg a \,\&\, b, \quad a \,\&\, \neg b, \quad \neg a \,\&\, \neg b, \text{ and } \neg a \vee \neg b.$$

In commonsense reasoning, however, we only use the named operations. Why?

Maybe it is the question of efficiency? Maybe it is the question of efficiency? To check on this, we can use the experience of computer design, where the constituent binary gates are selected so as to make computations more efficient.

Unfortunately, this leads to a completely different set of binary operations: e.g., computers typically use "nand" gates, that implement the function $f(a, b) = \neg(a \,\&\, b) = \neg a \vee \neg b$, but they never use gates corresponding to implication. So, the usual selection of binary logical operations remains a mystery.

What we do in this chapter. In this chapter, we show that the usual choice of logical operations in the 2-valued logic can be explained by ... fuzzy logic.

33.2 Our Explanation

Why fuzzy logic. We are interested in operations in 2-valued logics, so why should we take fuzzy logic into account? The reason is straightforward: in commonsense reasoning, we deal not only with precisely defined statements, but also with imprecise ("fuzzy") ones. For example:

- We can say that the age of a person is 18 or above *and* and this person is a US citizen, so he or she is eligible to vote.
- We can also say that a student is good academically and enthusiastic about research, so this student can be recommended for the graduate school.

We researchers may immediately see the difference between these two uses of "and": precisely defined ("crisp") in the first case, fuzzy in the second case. However, to many people, these two examples are very similar.

So, to understand why some binary operations are used in commonsense reasoning and some are not, it is desirable to consider the use of each operation not only in the 2-valued logic, but also in the more general fuzzy case; see, e.g., [2–7].

Which fuzzy generalizations of binary operations should we consider? In fuzzy logic, in addition to value 1 (true) and 0 (false), we also consider intermediate values corresponding to uncertainty. To describe such intermediate degrees, it is reasonable to consider real numbers intermediate between 0 or 1—this was exactly the original Zadeh's idea which is still actively used in applications of fuzzy logic.

So, to compare different binary operations, we need to extend these operations from the original set $\{0, 1\}$ to the whole interval $[0, 1]$. There are many possible extension of this type; which one should we select?

A natural idea is to select the most robust operations. For each fuzzy statement, its degree of confidence has to be elicited from the person making this statement. These statements are fuzzy, so naturally, it is not reasonable to expect that the same expert will always produce the exact same number: for the same statement, the expert can one day produce one number, another day a slightly different number, etc. When we plot these numbers, we will get something like a bell-shaped histogram—similar to what we get when we repeatedly measure the same quantity by the same measuring instrument. It is therefore reasonable to say that, in effect, the value a marked by an expert can differ from the corresponding mean value $\bar{a}$ by some small random value Δa, with zero mean and small standard deviation σ.

Since this small difference does not affect the user's perception, it should not affect the result of commonsense reasoning—in particular, the result of applying a binary operation to the corresponding imprecise numbers should not change much if we use slightly different estimates of the same expert. For example, 0.9 and 0.91 probably represent the same degree of expert's confidence. So, it is not reasonable to expect that we should get drastically different values of $a \& b$ if we use $a = 0.9$ or $a = 0.91$.

To be more precise, if we fix the value of one of the variables in a binary operation, then the effect of changing the second value on the result should be as small as possible. Due to the probabilistic character, we can only talk about being small "on average", i.e., about the smallest possible mean square difference. This idea was, in effect, presented in [5, 8, 9] for the case of "and"- and "or"-operations; let us show how it can be extended to all possible binary logical operations.

For a function $F(a)$ of one variable, if we replace a with $a + \Delta a$, then the value $F(a)$ changes to $F(a + \Delta a)$. Since the difference Δa is small, we can expand the above expression in Taylor series and ignore terms which are quadratic (or higher order) in terms of Δa. Thus, we keep only linear terms in this expansion: $F(a + \Delta a) = F(a) + F'(a) \cdot \Delta a$, where $F'(a)$, as usual, indicates the derivative. The resulting difference in the value of $F(a)$ is equal to $\Delta F \stackrel{\text{def}}{=} F(a + \Delta a) -$

$F(a) = F'(a) \cdot \Delta a$. Here, the mean squared value (variance) of Δa is equal to σ^2; thus, the mean squared value of $F'(a) \cdot \Delta a$ is equal to

$$(F'(a))^2 \cdot \sigma^2.$$

We are interested in the mean value of this difference. Here, a can take any value from the interval $[0, 1]$, so the resulting mean value takes the form $\int_0^1 (F'(a))^2 \cdot \sigma^2 \, da$. Since σ is a constant, minimizing this difference is equivalent to minimizing the integral $\int_0^1 (F'(a))^2 \, da$.

Our goal is to extend a binary operation from the 2-valued set $\{0, 1\}$. So, we usually know the values of the function $F(a)$ for $a = 0$ and $a = 1$. In general, a function $f(x)$ that minimizes a functional $\int L(f, f') \, dx$ is described by the Euler-Lagrange equation

$$\frac{\partial L}{\partial f} - \frac{d}{dx} \frac{\partial L}{\partial f'} = 0;$$

see, e.g., [10]. In our case, $L = (F'(a))^2$, so the Euler-Lagrange equation has the form

$$\frac{d}{da}(2F'(a)) = 2F''(a) = 0.$$

Thus, $F''(a) = 0$, meaning that the function $F(a)$ is linear.

Thus, we conclude that in our extensions of binary (and other) operations, the corresponding function should be linear in each of the variables, i.e., that this function should be bilinear (or, in general, multi-linear).

Such an extension is unique: a proof. Let us show that this requirement of bilinearity uniquely determines the corresponding extension. Indeed, suppose that two bilinear expressions $f(a, b)$ and $g(a, b)$ have the same values when $a \in \{0, 1\}$ and $b \in \{0, 1\}$. In this case, the difference $d(a, b) = f(a, b) - g(a, b)$ is also a bilinear expression whose value for all four pairs (a, b) for which $a \in \{0, 1\}$ and $b \in \{0, 1\}$ is equal to 0. Since $d(0, 0) = d(1, 0) = 0$, by linearity, we conclude that $d(a, 0) = 0$ for all a. Similarly, since $d(0, 1) = d(1, 1) = 0$, by linearity, we conclude that $d(a, 1) = 0$ for all a. Now, since $d(a, 0) = d(a, 1) = 0$, by linearity, we conclude that $d(a, b) = 0$ for all a and b—thus, indeed, $f(a, b) = g(a, b)$ and the extension is indeed defined uniquely.

How can we find the corresponding bilinear extension. For negation, linear interpolation leads to the usual formula $f(a) = 1 - a$.

Let us see what happens for binary operations. Once we know the values $f(0, 0)$ and $f(1, 0)$, we can use linear interpolation to find the values $f(a, 0)$ for all a:

$$f(a, 0) = f(0, 0) + a \cdot (f(1, 0) - f(0, 0)).$$

Similarly, once we know the values $f(0, 1)$ and $f(1, 1)$, we can use linear interpolation to find the values $f(a, 1)$ for all a:

$$f(a, 1) = f(0, 1) + a \cdot (f(1, 1) - f(0, 1)).$$

Now, since we know, for each a, the values $f(a, 0)$ and $f(a, 1)$, we can use linear interpolation to find the value $f(a, b)$ for each b as

$$f(a, b) = f(a, 0) + b \cdot (f(a, 1) - f(a, 0)).$$

Let us use this idea to find the bilinear extensions of all ten non-trivial binary operations.

Let us list bilinear extensions of all non-trivial binary operations.

- for $a \& b$, we have $f(a, b) = a \cdot b$;
- for $a \& \neg b$, we have $f(a, b) = a \cdot (1 - b) = a - a \cdot b$;
- for $\neg a \& b$, we have $f(a, b) = (1 - a) \cdot b = b - a \cdot b$;
- for $a \oplus b$, we have $f(a, b) = a + b - 2a \cdot b$;
- for $a \vee b$, we have $f(a, b) = a + b - a \cdot b$;
- for $\neg a \& \neg b$, we have $f(a, b) = (1 - a) \cdot (1 - b)$;
- for $a \equiv b$, we have $f(a, b) = 1 - a - b + 2a \cdot b$;
- for $\neg a \vee b = a \rightarrow b$, we have $f(a, b) = 1 - a + a \cdot b$;
- for $\neg a \vee b = b \rightarrow a$, we have $f(a, b) = 1 - b + a \cdot b$;
- finally, for $\neg a \vee \neg b = \neg(a \& b)$, we have $f(a, b) = 1 - a \cdot b$.

Which operations should we select as basic. As the basic operations, we should select the ones which are the easiest to compute. Of course, we should have *negation* $f(a) = 1 - a$, since it is the easiest to compute: it requires only one subtraction.

Out of all the above ten binary operations, the simplest to compute is $f(a, b) = a \cdot b$ (corresponding to "*and*") which requires only one arithmetic operation—multiplication. All other operations need at least one more arithmetic operation. This explains why "and" is one of the basic operations in commonsense reasoning.

Several other operations can be described in terms of the selected "and"- and "not"-operations $f_\&(a, b) = a \cdot b$ and $f_\neg(a) = 1 - a$:

- the operation $f(a, b) = a - a \cdot b$ corresponding to $a \& \neg b$ can be represented as $f_\&(a, f_\neg(b))$;
- the operation $f(a, b) = b - a \cdot b$ corresponding to $\neg a \& b$ can be represented as $f_\&(f_\neg(a), b)$;
- the operation $f(a, b) = a + b - a \cdot b$ corresponding to $a \vee b$ can be represented as $f_\neg(f_\&(f_\neg(a), f_\neg(b)))$;
- the operation $f(a, b) = (1 - a) \cdot (1 - b)$ corresponding to $\neg a \& \neg b$ can be represented as $f_\&(f_\neg(a), f_\neg(b))$;
- the operation $f(a, b) = 1 - a + a \cdot b$ corresponding to $\neg a \vee b = a \rightarrow b$ can be represented as $f_\neg(f_\&(a, f_\neg(b)))$;
- the operation $f(a, b) = 1 - b + a \cdot b$ corresponding to $a \vee \neg b = b \rightarrow a$ can be represented as $f_\neg(f_\&(f_\neg(a), b))$; and
- the operation $f(a, b) = 1 - a \cdot b$ corresponding to $\neg a \vee \neg b = \neg(a \& b)$ can be represented as $f_\neg(f_\&(a, b))$.

There are two binary operations which *cannot* be represented as compositions of the selected "and" and "or"-operations: namely:

- the operation $f(a, b) = a + b - 2a \cdot b$ corresponding to exclusive "or" $a \oplus b$, and
- the operation $f(a, b) = 1 - a - b + 2a \cdot b$ corresponding to equivalence $a \equiv b$.

Thus, we need to add at least one of these operations to our list of basic operations. Out of these two operations, the simplest to compute—i.e., requiring the smallest number of arithmetic operations—is the function corresponding to *exclusive "or"*. This explains why we use exclusive "or" in commonsense reasoning.

What about operations that *can* be described in terms of "and" and "not"? For some of these operations, e.g., for the function

$$f(a, b) = a - a \cdot b = a \cdot (1 - b)$$

corresponding to $a \& \neg b$, direct computation requires exactly as many arithmetic operations as computing the corresponding representation in terms of "and" and "or". However, there is one major exception: for the function $f(a, b) = a + b - a \cdot b$ corresponding to "or":

- its straightforward computation requires two additions/subtractions and one multiplication, while
- its computation as $f_\neg(f_\&(f_\neg(a), f_\neg(b)))$ requires one multiplication (when applying $f_\&$) but *three* subtractions (corresponding to three negation operations).

Thus, for the purpose of efficiency, it makes sense to consider *"or"* as a separate operation. This explains why we use "or" in commonsense reasoning.

So far, we have explained all basic operations except for implication. Explaining implication requires a slightly more subtle analysis.

Why implications. In the above analysis, we considered addition and subtraction to be equally complex. This is indeed the case for computer-based computations, but for us humans, subtraction is slightly more complex than addition. This does not change our conclusion about operations like $f(a, b) = a - a \cdot b$: whether we compute them directly or as $f(a, b) = f_\&(a, f_\neg(b))$, in both cases, we use the same number of multiplications, the same number of additions, and the same number of subtractions.

There is, however, a difference for implication operations such as

$$f(a, b) = 1 - a + a \cdot b = f_\neg(f_\&(a, f_\neg(b))) :$$

- its direct computation requires one multiplication, one addition, and one subtraction, while
- its computation in terms of "and"- and "not"-operations requires one multiplication and two subtractions.

In this sense, the direct computation of implication is more efficient—which explains why we also use implication in commonsense reasoning.

Conclusion. By using fuzzy logic, we have explained why negation, "and", "or", implication, and exclusive "or" are used in commonsense reasoning while other binary 2-valued logical operations are not.

33.3   Auxiliary Result: Why the Usual Quantifiers?

Formulation of the problem. In the previous sections, we consider binary logical operations. In our reasoning, we also use quantifiers such as "for all" and "there exists" which are, in effect, n-ary logical operations, where n is the number of possible objects.

Why these quantifiers? Why not use additional quantifiers like "there exists at least two"? Let us analyze this question from the same fuzzy-based viewpoint from which we analyzed binary operations. It turns out that this way, we get a (partial) explanation for the usual choice of quantifiers.

Why universal quantifier. Let us consider the case of n objects $1, \ldots, n$. We have some property $p(i)$ which, for each object i, can be true or false. We would like to combine these n truth values into a single one, i.e., we need an n-ary operation $f(a_1, \ldots, a_n)$ that would transform n truth values $p(1), \ldots, p(n)$ into a single combined value $f(p(1), \ldots, p(n))$.

Similarly to the previous chapter, let us consider a fuzzy version. Just like all non-degenerate binary operations—i.e., operations that are not constants or unary operations—must contain a product of two numbers, similarly, all non-degenerate n-ary operations must contain the product of n numbers.

Thus, the simplest possible case—with the fastest computations—is when the operation is simply the product of the given n numbers, i.e., the operation $f(a_1, \ldots, a_n) = a_1 \cdot \ldots \cdot a_n$. Indeed, every other operation requires also addition of subtraction. This operation transforms the values $p(1), \ldots, p(n)$ into their product $p(1) \cdot \ldots \cdot p(n)$, that corresponds exactly to the formula $p(1) \& \ldots \& p(n)$, i.e., to the formula $\forall i \; p(i)$. This explains the ubiquity of the universal quantifiers.

Why existential quantifier. The universal quantifier has the property that it does not change if we permute the objects: $\prod_{i=1}^{n} p(i) = \prod_{i=1}^{n} p(\pi(i))$ for every permutation $\pi : \{1, \ldots, n\} \to \{1, \ldots, n\}$. This condition of permutation invariance holds for all quantifiers and it is natural to be required. We did not explicitly impose this condition in our derivation of universal quantifier for only one reason—that we were able to derive this quantifier only from the requirement of computational simplicity, without a need to also explicitly require permutation invariance.

However, now that we go from the justification of the simplest possible quantifier to a justification of other quantifiers, we need to explicitly require permutation invariance—otherwise, the next simplest operations are operations like

$$\neg a_1 \,\&\, a_2 \,\&\, a_2 \,\ldots\, \&\, a_n.$$

It turns out that among permutation-invariant n-ary logical operations, the simplest are:

- the operation $\neg\forall i\ p(i)$ for which the corresponding formula $1 - \prod_{i=1}^{n} p_i$ requires $n - 1$ multiplications and one subtraction;
- the operation $\forall i\ \neg p(i)$ for which the corresponding formula $\prod_{i=1}^{n}(1 - p_i)$ requires $n - 1$ multiplications and n subtractions; and
- the operation $\exists i\ p(i)$, i.e., equivalently, $\neg\forall i\ \neg p(i)$, for which the corresponding formula $1 - \prod_{i=1}^{n}(1 - p_i)$ requires $n - 1$ multiplications and $n + 1$ subtractions.

This explains the ubiquity of existential quantifiers.

References

1. Urenda, J., Kosheleva, O., Kreinovich, V.: Fuzzy logic explains the usual choice of logical operations in 2-valued logic. In: Ceberio, M., Kreinovich, V. (eds.) How Uncertainty-Related Ideas Can Provide Theoretical Explanation for Empirical Dependencies, pp. 141–151. Springer, Cham, Switzerland (2021)
2. Belohlavek, R., Dauben, J.W., Klir, G.J.: Fuzzy Logic and Mathematics: A Historical Perspective. Oxford University Press, New York (2017)
3. Klir, G., Yuan, B.: Fuzzy Sets and Fuzzy Logic. Prentice Hall, Upper Saddle River, New Jersey (1995)
4. Mendel, J.M.: Uncertain Rule-Based Fuzzy Systems: Introduction and New Directions. Springer, Cham, Switzerland (2017)
5. Nguyen, H.T., Walker, C.L., Walker, E.A.: A First Course in Fuzzy Logic. Chapman and Hall/CRC, Boca Raton, Florida (2019)
6. Novák, V., Perfilieva, I., Močkoř, J.: Mathematical Principles of Fuzzy Logic. Kluwer, Boston, Dordrecht (1999)
7. Zadeh, L.A.: Fuzzy sets. Inf. Control **8**, 338–353 (1965)
8. Nguyen, H.T., Kreinovich, V., Tolbert, D.: On robustness of fuzzy logics. In: Proceedings of the 1993 IEEE International Conference on Fuzzy Systems FUZZ-IEEE'93, vol. 1, pp. 543–547. San Francisco, California (1993)
9. Tolbert, D.: Finding "and" and "or" operations that are least sensitive to change in intelligent control, University of Texas at El Paso, Department of Computer Science, Master's Thesis (1994)
10. Gelfand, I.M., Fomin, S.V.: Calculus of Variations. Dover Publication, New York (2000)

Chapter 34
Data Processing: Probabilistic Techniques, Part 1

A known alternative to fuzzy techniques are probabilistic techniques. They work perfectly well when we have the full information about the corresponding probabilities, but in many practical situations, we do not have this information. In this case, as we show in this chapter and in the following chapter, algebraic approach can help select the corresponding probability distributions.

In this chapter, we take into account that in many practical situations, we only know the interval containing the quantity of interest, we have no information about the probability of different values within this interval. In contrast to the cases when we know the distributions and can thus use Monte-Carlo simulations, processing such interval uncertainty is difficult—crudely speaking, because we need to try all possible distributions on this interval. Sometimes, the problem can be simplified: namely, it is possible to select a single distribution (or a small family of distributions) whose analysis provides a good understanding of the situation. The most known case is when we use the Maximum Entropy approach and get the uniform distribution on the interval. Interesting, sensitivity analysis—which has completely different objectives—leads to selection of the same uniform distribution. In this chapter, we provide a general explanation of why uniform distribution appears in different situations—namely, it appears every time we have a permutation-invariant objective functions with the unique optimum. We also discuss what happens if there are several optima.

Results from this chapter first appeared in [1].

34.1 Problem: How to Represent Interval Uncertainty

Interval uncertainty is ubiquitous. When an engineer designs an object, the original design comes with exact numerical values of the corresponding quantities, be it the height of ceiling in civil engineering or the resistance of a certain resistor in electrical engineering. Of course, in practice, it is not realistic to maintain the exact values of all these quantities, we can only maintain them with some tolerance. As a result, the engineers not only produce the desired ("nominal") value x of the corresponding quantity, they also provide a tolerance $\varepsilon > 0$ with which we need to maintain the value of this quantity. The actual value must be in the interval $\mathbf{x} = [\underline{x}, \overline{x}]$, where $\underline{x} \stackrel{\text{def}}{=} x - \varepsilon$ and $\overline{x} \stackrel{\text{def}}{=} x + \varepsilon$.

All the manufacturers need to do is to follow these interval recommendations. There is no special restriction on probabilities of different values within these intervals—these probabilities depends on the manufacturer, and even for the same manufacturer, they may change every time the manufacturer makes some adjustments to the manufacturing process.

Data processing under interval uncertainty is often difficult. Because of the ubiquity of interval uncertainty, many researchers have considered different data processing problems under this uncertainty; this research area is known as *interval computations*; see, e.g., [2–5].

The problem is that the corresponding computational problems are often very complex, much more complex than solving similar problems under *probabilistic* uncertainty—when we know the probabilities of different values within the corresponding intervals. For example, while for the probabilistic uncertainty, we can, in principle, always use Monte-Carlo simulations to understand how the input uncertainty affects the result of data processing, a similar problem for interval uncertainty is NP-hard already for the simplest nonlinear case when the whole data processing means computing the value of a quadratic function—actually, it is even NP-hard if we want to find the range of possible values of variance in a situation when inputs are only known with interval uncertainty [6, 7].

This complexity is easy to understand: interval uncertainty means that we may have different probability distributions on the given interval. So, to get guaranteed estimates, we need to consider all of them—which leads to very time-consuming computations. For some problems, this time can be sped up, but in general, the problems remain difficult.

It is desirable to have a reasonably small family of distributions representing interval uncertainty. Considering all possible distributions on an interval will take forever. It is therefore desirable to look for cases when interval uncertainty can be represented by a single distribution—or at least by a reasonably small family of distributions, e.g., by finitely many distributions or by a finite-dimensional family.

Maximum entropy idea. In general, the problem of selecting a single distribution from the family of all distributions which are consistent with our knowledge (i.e., with measurement results and known general principles) is well known in data processing. A usual solution to this problem is to select the distribution which best represents the corresponding uncertainty. Some possible distributions have little uncertainty—e.g., we can have a distribution which is located at some point with probability 1. Selecting such a distribution would mislead the data processing algorithm into thinking that we have no uncertainty at all. Similarly, selecting a distribution which is located on a proper subinterval of the original interval would be misleading—since it will lead to a decrease in perceived uncertainty. From this viewpoint, a proper selection should select a representative distribution with the largest possible uncertainty.

A reasonable measure of uncertainty is *entropy*

$$S = - \int \rho(x) \cdot \ln(\rho(x)) \, dx, \tag{34.1}$$

where $\rho(x)$ denotes the probability density function; see, e.g., [7, 8]. So, a reasonable idea is to select, among all possible distributions, a distribution with the largest possible entropy. This idea is known as the *Maximum Entropy* approach, and it has indeed very successful in many applications; see, e.g., [8].

For interval uncertainty, maximum entropy leads to the uniform distribution. What do we get when we apply the maximum entropy approach to the case of interval uncertainty, when all we know is that the probability distribution is located in some interval $[a, b]$? In this case, we want to find a distribution $\rho(x)$ for which the entropy (34.1) is maximized under the condition that the overall probability is 1, i.e., that $\int_a^b \rho(x) \, dx = 1$.

The usual way to solve such constraint optimization problems is to use the Lagrange multiplier method, where the problem of optimizing a function $f(A)$ under the constraint $g(A) = 0$ is reduced to an unconstrained problem of optimizing the auxiliary function $f(A) + \lambda \cdot g(A)$, with the parameter λ (known as the *Lagrange multiplier*) to be determined from the condition that the resulting optimizing alternative A satisfy the original constraint.

In our case, this means that we maximize the auxiliary function

$$- \int \rho(x) \cdot \ln(\rho(x)) \, dx + \lambda \cdot \left(\int_a^b \rho(x) \, dx - 1 \right).$$

Strictly speaking, this expression has infinitely many unknowns—namely, the values $\rho(x)$ corresponding to all possible values x. However, in practice, we can always take into account that even with the best possible measuring instruments, we can only measure the value of the physical quantity x with some uncertainty h. Thus, from the practical viewpoint, it makes sense to divide the interval $[a, b]$ into small subintervals $[a, a + h], [a + h, a + 2h], \ldots$ within each of which the values of x are indistinguishable, and instead of the function $\rho(x)$, consider the probabilities $p_1, p_2, \ldots$ of the value x being in each of these intervals. In these terms, the entropy takes the form $S = - \sum_{i=1}^{n} p_i \cdot \ln(p_i)$, the requirement that probabilities add up to 1 take the form $\sum_{i=1}^{n} p_i = 1$, and the resulting equivalent unconstrained optimization problem takes the form of maximizing the expression

$$-\sum_{i=1}^{n} p_i \cdot \ln(p_i) + \lambda \cdot \left(\sum_{i=1}^{n} p_i - 1\right).$$

To find the maximum value of this expression, the usual idea is to differentiate this expression with respect to each unknown p_i and equate the resulting derivative to 0. As a result, we get the formula $-\ln(p_i) - 1 + \lambda = 0$, hence $\ln(p_i) = 1 + \lambda$, and $p_i = \exp(1 + \lambda)$.

This value is the same for all i. Thus, the probability to be in each of the small subintervals is the same—i.e., we select a uniform distribution on the original interval $[0, 1]$.

Need to go beyond the maximum entropy approach: first argument. In different practical problems, we can have different objective functions. For example, in many cases, it is important to know how sensitive is the system to different perturbations. In this case, it is also desirable to select one distribution (or at least a reasonably small family of distributions).

The corresponding objective function is very different from the entropy. However, interestingly, it turned out that the best choice of a representative distribution is still a uniform distribution on the given interval; see, e.g., [9]. Why?

The fact that two different optimization problems lead to the exact same selection makes us think that there must be a fundamental reason behind these two results—and in this chapter, we indeed describe such a reason.

Need to go beyond the maximum entropy approach: second argument. When we select a single distribution, we can find the *largest* possible entropy of the corresponding distributions. But what if, in addition to the largest possible value of the entropy, we want to know the whole *range* of values of entropy—i.e., we also want to know the *smallest* possible value?

This smallest possible value is not attained on a single probability distribution—in the discrete case, it is attained for all distributions for which $p_{i_0} = 1$ for some i_0 and for which $p_i = 0$ for all $i \neq i_0$. In this case, we cannot select a single distribution—the maximum entropy approach does not help, but we can still select a small representative *family* of distributions.

Again, a natural question is: can we generalize this result so that it would cover other practically useful situations?

34.2   Analysis of the Problem

What do entropy and sensitivity measure have in common? We would like to come up with a general result that generalizes both the maximum entropy and the sensitivity results. To come up with such a generalization, it is reasonable to analyze what these two results have in common.

Let us use symmetries. In general, our knowledge is based on *symmetries*, i.e., on the fact that some situations are similar to each other. Indeed, it all the world's situations were completely different, we would not be able to make any predictions. Luckily, real-life situations have many features in common, so we can use the experience of previous situations to predict future ones.

For example, when a person drops a pen, it starts falling down to Earth with the acceleration of 9.81 m/s². If this person moves to a different location and repeats the same experiment, he or she will get the exact same result. This means that the corresponding physics is invariant with respect to shifts in space.

Similarly, if the person repeats this experiment in a year, the result will be the same. This means that the corresponding physics is invariant with respect to shifts in time.

Alternatively, if the person turns around a little bit, the result will still be the same. This means that the underlying physics is also invariant with respect to rotations, etc.

This is a very simple example, but such symmetries are invariances are actively used in modern physics (see, e.g., [10, 11])—and moreover, many previously proposed fundamental physical theories such as:

- Maxwell's equations that describe electrodynamics,
- Schroedinger's equations that describe quantum phenomena,
- Einstein's General Relativity equation that describe gravity,

can be derived from the corresponding invariance assumptions; see, e.g., [12–15].

Symmetries also help to explain many empirical phenomena in computing; see, e.g., [16]. From this viewpoint, a natural way to look for what the two examples have in common is to look for invariances that they have in common.

Permutations—natural symmetries in the entropy example. We have n probabilities $p_1, \ldots, p_n$. What can we do with them that would preserve the entropy? In principle, we can transform the values into something else, but the easiest possible transformations is when we do not change the values themselves, just swap them.

Bingo! Under such swap, the value of the entropy does not change. In precise terms, both the objective function $S = - \sum_{i=1}^{n} p_i \cdot \ln(p_i)$ and the constraint $\sum_{i=1}^{n} p_i = 1$ do not change is we perform any permutation

$$\pi : \{1, \ldots, n\} \to \{1, \ldots, n\},$$

i.e., replace the values $p_1, \ldots, p_n$ with the permuted values $p_{\pi(1)}, \ldots, p_{\pi(n)}$.

Permutations also work for the sensitivity example. Interestingly, a more complex criterion used in the sensitivity example is also permutation-invariant: as well as many other generalization of entropy.

Thus, we are ready to present our general results.

34.3 Our Results

Definition 34.1 • We say that a function $f(p_1, \ldots, p_n)$ is *permutation-invariant* if for every permutation $\pi : \{1, \ldots, n\} \to \{1, \ldots, n\}$, we have

$$f(p_1, \ldots, p_n) = f(p_{\pi(1)}, \ldots, p_{\pi(n)}).$$

• By a permutation-invariant optimization problem, we mean a problem of optimizing a permutation-invariant function $f(p_1, \ldots, p_n)$ under constraints of the type $g(p_1, \ldots, p_n) = a$ or $g(p_1, \ldots, p_n) \geq a$ for permutation-invariant functions g.

Proposition 34.1 *If a permutation-invariant optimization problem has only one solution, then for this solution, we have $p_1 = \cdots = p_n$.*

Discussion. This explains why we get the uniform distribution both in the maximum entropy case and in the sensitivity case.

Proof We will prove by contradiction. Suppose that the values p_i are not all equal. This means that there exist i and j for which $p_i \neq p_j$. Let us swap p_i and p_j, and denote the corresponding values by p'_i, i.e.:

• we have $p'_i = p_j$,
• we have $p'_j = p_i$, and
• we have $p'_k = p_k$ for all other k.

Since the values p_i satisfy all the constraints, and all the constraints are permutation-invariant, the new values p'_i also satisfy all the constraints. Since the objective function is permutation-invariant, we have $f(p_1, \ldots, p_n) = f(p'_1, \ldots, p'_n)$. Since the values $(p_1, \ldots, p_n)$ were optimal, the values $(p'_1, \ldots, p'_n) \neq (p_1, \ldots, p_n)$ are thus also optimal—which contradicts to the assumption that the original problem has only one solution.

This contradiction proves for the optimal tuple $(p_1, \ldots, p_n)$ that all the values p_i are indeed equal to each other. The proposition is proven.

Discussion. What is the optimal solution is not unique? We can have a case when we have a small finite number of solutions.

We can also have a case when we have a 1-parametric family of solutions—i.e., a family depending on one parameter. In our discretized formulation, each parameter has n values, so this means that we have n possible solutions. Similarly, a 2-parametric family means that we have n^2 possible solutions, etc.

Here are precise definitions and related results.

Definition 34.2 Let the number n of variable p_i be fixed.

• We say that a problem has a *small finite number of solutions* if its number of solutions is smaller than n.
• We say that a problem has a *d-parametric family of solutions* if it has n^d solutions.

Proposition 34.2 *If a permutation-invariant optimization problem has a small finite number solutions with $\sum p_i = 1$, then it has only one solution.*

Discussion. Due to Proposition 34.1, in this case, the only solution is the uniform distribution $p_1 = \cdots = p_n$.

Proof Since $\sum p_i = 1$, there is only one possible solution for which $p_1 = \cdots = p_n$: the solution for which all the values p_i are equal to $1/n$.

Thus, if the problem has more than one solution, some values p_i are different. Let us pick one such value p_k. Let S denote the set of all the indices j for which $p_j = p_k$, and let m denote the number of elements in this set. Since some values p_i are different, we have $1 \leq m \leq n - 1$.

Due to permutation-invariance, each permutation of this solution is also a solution. For each m-size subset of the set of n-element set of indices $\{1, \ldots, n\}$, we can have a permutation that transforms S into this set and thus, produces a new solution to the original problem. There are $\binom{n}{m}$ such subsets. For all m from 1 to $n - 1$, the smallest value of the binomial coefficient $\binom{n}{m}$ is attained when $m = 1$ or $m = n - 1$, and this smallest value is equal to n. Thus, if there is more than one solution, we have at least n different solutions—and since we assumed that we have fewer than n solutions, this means that we have only one. The proposition is proven.

Proposition 34.3 *If a permutation-invariant optimization problem has a 1-parametric family of solutions $p_i \geq 0$ with $\sum p_i = 1$, then this family is characterized by a real number $c \leq 1/(n - 1)$, for which all these solutions have the following form: $p_i = c$ for all i but one and $p_{i_0} = 1 - (n - 1) \cdot c$ for the remaining value i_0.*

Discussion. In particular, for $c = 0$, we get the above-mentioned 1-parametric family of distributions for which entropy is the smallest possible.

Proof As we have shown in the proof of Proposition 34.2, if in one of the solutions, for some value p_i we have m different indices j with this value, then we will have at least $\binom{n}{m}$ different solutions. For all m from 2 to $n - 2$, this number is at least as large as $\binom{n}{2} = \dfrac{n \cdot (n - 1)}{2}$ and is, thus, larger than n.

Since overall, we only have n solutions, this means that it is not possible to have $2 \leq m \leq n - 2$. So, the only possible values of m are 1 and $n - 1$.

If there was no group with $n - 1$ values, this would means that all the groups must have $m = 1$, i.e., consist of only one value. In other words, in this case, all n values p_i would be different. In this case, each of $n!$ permutations would lead to a different solution—so we would have $n! > n$ solutions to the original problem—but we assumed that overall, there are only n solutions. Thus, this case is also impossible.

So, we do have a group of $n - 1$ values with the same p_i. Then we get exactly one of the solutions described in the formulation of the proposal, plus solutions obtained from it by permutations—which is exactly the described family.

The proposition is proven.

References

1. Beer, M., Urenda, J., Kosheleva, O., Kreinovich, V.: Which distributions (or families of distributions) best represent interval uncertainty: case of permutation-invariant criteria. In: Proceedings of the 18th International Conference on Information Processing and Management of Uncertainty in Knowledge-Based Systems IPMU'2020, Lisbon, Portugal, 5–19 June, pp. 70–79 (2020)
2. Jaulin, L., Kiefer, M., Didrit, O., Walter, E.: Applied Interval Analysis, with Examples in Parameter and State Estimation, Robust Control, and Robotics. Springer, London (2001)
3. Mayer, G.: Interval Analysis and Automatic Result Verification. de Gruyter, Berlin (2017)
4. Moore, R.E., Kearfott, R.B., Cloud, M.J.: Introduction to Interval Analysis. SIAM, Philadelphia (2009)
5. Rabinovich, S.G.: Measurement Errors and Uncertainties: Theory and Practice. Springer, New York (2005)
6. Kreinovich, V., Lakeyev, A., Rohn, J., Kahl, P.: Computational Complexity and Feasibility of Data Processing and Interval Computations. Kluwer, Dordrecht (1998)
7. Nguyen, H.T., Kreinovich, V., Wu, B., Xiang, G.: Computing Statistics under Interval and Fuzzy Uncertainty. Springer, Berlin (2012)
8. Jaynes, E.T., Bretthorst, G.L.: Probability Theory: The Logic of Science. Cambridge University Press, Cambridge (2003)
9. He, L., Beer, M., Broggi, M., Wei, P., Gomes, A.T.: Sensitivity analysis of prior beliefs in advanced Bayesian networks. In: Proceedings of the 2019 IEEE Symposium Series on Computational Intelligence SSCI'2019, Xiamen, China, December 6–9, pp. 775–782 (2019)
10. Feynman, R., Leighton, R., Sands, M.: The Feynman Lectures on Physics. Addison Wesley, Boston (2005)
11. Thorne, K.S., Blandford, R.D.: Modern Classical Physics: Optics, Fluids, Plasmas, Elasticity, Relativity, and Statistical Physics. Princeton University Press, Princeton, New Jersey (2017)
12. Finkelstein, A.M., Kreinovich, V.: Derivation of Einstein's, Brans-Dicke and other equations from group considerations. In: Choque-Bruhat, Y., Karade, T.M. (eds), On Relativity Theory. Proceedings of the Sir Arthur Eddington Centenary Symposium, Nagpur India 1984, vol. 2, pp. 138–146. World Scientific, Singapore (1985)
13. Finkelstein, A.M., Kreinovich, V., Zapatrin, R.R.: Fundamental physical equations uniquely determined by their symmetry groups. In: Lecture Notes in Mathematics, vol. 1214, pp. 159–170. Springer, Berlin (1986)
14. Kreinovich, V.: Derivation of the Schroedinger equations from scale invariance. Theor. Math. Phys. **8**(3), 282–285 (1976)
15. Kreinovich, V., Liu, G.: We live in the best of possible worlds: Leibniz's insight helps to derive equations of modern physics. In: Pisano, R., Fichant, M., Bussotti, P., Oliveira, A.R.E. (eds.), The Dialogue between Sciences, Philosophy and Engineering. New Historical and Epistemological Insights, Homage to Gottfried W. Leibnitz 1646–1716, pp. 207–226. College Publications, London (2017)
16. Nguyen, H.T., Kreinovich, V.: Applications of Continuous Mathematics to Computer Science. Kluwer, Dordrecht (1997)

Chapter 35
Data Processing: Probabilistic Techniques, Part 2

In this chapter, we will show how to use probability distributions in other situations in which we do not have fuzzy information about the probabilities.

As a consequence of our result, we will also get an explanation for the following puzzling empirical fact. In principle, any non-negative function can serve as a probability density function—provided that it adds up to 1. All kinds of processes are possible, so it seems reasonable to expect that observed probability density functions are random with respect to some appropriate probability measure on the set of all such functions—and for all such measures, similarly to the simplest case of random walk, almost all functions have infinitely many local maxima and minima. However, in practice, most empirical distributions have only a few local maxima and minima—often one (unimodal distribution), sometimes two (bimodal), and, in general, they are few-modal. From this viewpoint, econometrics is no exception: empirical distributions of economics-related quantities are also usually few-modal. In this chapter, we provide a theoretical explanation for this empirical fact.

Results from this chapter first appeared in [1].

35.1 Problem: How to Represent General Uncertainty

Main idea. Of course, the space of all possible probability density functions is infinite-dimensional, so to exactly describe each such function, we need to describe the values of infinitely many parameters. In practice, at each moment of time, we can only use finitely many parameters. So, we need to look into appropriate finite-dimensional families of probability density functions—and explain why functions from this appropriate family are few-modal.

To answer this question, let us describe natural properties of such families F of distributions $\rho(c_1, \ldots, c_n, x)$. To come up with these properties, let us recall how we gain the information about the corresponding distributions.

We want smoothness. Small changes in the values of the parameters c_i and/or small changes in x should lead to small changes in the probability density. In other words, we want the function $\rho(c_1, \ldots, c_n, x)$ to be smooth.

We can combine different pieces of knowledge. Suppose that:

- one piece of evidence leads us to conclude that the distribution of the corresponding quantity is described by a probability density function $\rho_1(x)$, and
- another—independent—piece of evidence—leads to a slightly different probability density function $\rho_2(x)$.

If these were evidences about two different quantities x_1 and x_2, then, due to independence, we would conclude that the distribution of the pair (x_1, x_2) follows a product distribution $\rho_1(x_1) \cdot \rho_2(x_2)$. In our case, however, we know that this is the same quantity, i.e., that $x_1 = x_2$. Thus, to get the resulting distribution, we need to restrict the product distribution to the case when $x_1 = x_2$, i.e., in precise terms, we need to consider conditional distribution under the condition that $x_1 = x_2$. This means that we need to consider the distribution $\rho(x) = c \cdot \rho_1(x) \cdot \rho_2(x)$, where c is a normalizing constant—which can be determined by the condition that $\int \rho(x)\, dx = 1$.

Thus, it is reasonable to require that for every two distribution $\rho_1(x)$ and $\rho_2(x)$ from the desired family F, their normalized product $c \cdot \rho_1(x) \cdot \rho_2(x)$ should also belongs to this family.

Knowledge can come in parts. Sometimes, we gain the knowledge right away. In many other cases, knowledge comes in small steps. If the resulting knowledge is described by a probability density function $\rho(x)$, and it comes via several (n) independent similar pieces of knowledge, each characterized by some probability density function $\rho_1(x)$, then, based on the previous subsection, we can conclude that $\rho(x) = c \cdot (\rho_1(x))^n$ for some constant c, i.e., that $\rho_1(x) = c_1 \cdot (\rho(x))^{1/n}$ for an appropriate normalizing coefficient c_1.

Thus, it is reasonable to require that for every distribution $\rho_1(x)$ from the desired family F and for every natural number $n > 1$, the normalized distribution $c_1 \cdot (\rho(x))^{1/n}$ should also belong to the family.

Scale- and shift-invariance. The numerical value of a quantity depends:

- on the starting point for measuring this quantity and
- on the measuring unit.

When we change numerical values, the expression for the probability distribution also changes. It is reasonable to require that if we simply change the starting point and/or the measuring unit in a distribution from the family F, then we should still get a distribution from the same family.

If we change the starting point, i.e., if we replace the original starting point with a new one which is a units larger, then in the new units $y = x - a$, the distribution described by the original probability density function $\rho(x)$ will now be described by the new function $\rho_1(y) = \rho(y + a)$.

Similarly, if we change the measuring unit, i.e., if we replace the original measuring unit with a new one which is λ times larger, then in the new units $y = x/\lambda$, the distribution described by the original probability density function $\rho(x)$ will now be described by the new function $\rho_1(y) = \lambda \cdot \rho(\lambda \cdot y)$.

Now, we are ready. Now, we are ready to formulate our main result.

35.2 Definitions and the Main Result

Definition 35.1 Let n be a natural number.

- By an *n-parametric family of distributions*, we mean a family

$$F = \{f(c_1, \ldots, c_n, x)\}_{c_1, \ldots, c_n}$$

of probability density functions, where the values $(c_1, \ldots, c_n)$ go over some set U, and the function $f(c_1, \ldots, c_n, x)$ is continuously differentiable over the closure of this set.
- We say that a family F *allows combining knowledge* if for very two functions $\rho_1(x)$ and $\rho_2(x)$ from this family, there exists a real number $c > 0$ for which the product $c \cdot \rho_1(x) \cdot \rho_2(x)$ also belongs to F.
- We say that a family F *allows partial knowledge* if for every function $\rho(x)$ from this family and for every natural number n, there exists a real number $c > 0$ for which the function $c \cdot (\rho(x))^{1/n}$ also belongs to F.
- We say that a family F is *shift-invariant* if for every function $\rho(x)$ from this family and for every real number a, the function $\rho(x + a)$ also belongs to F.
- We say that a family F is *scale-invariant* if for every function $\rho(x)$ from this family and for every real number $\lambda > 0$, the function $\lambda \cdot \rho(\lambda \cdot x)$ also belongs to F.

Proposition 35.1 *Every function from a shift- and scale-invariant n-parametric family of distributions that allows combining knowledge and partial knowledge has the form $\rho(x) = \exp(P(x))$ for some polynomial of degree $\leq n$.*

Proof of the main result.
$1°$. Let F be a family that satisfies all the given properties. To somewhat simplify the problem, let us consider a family G of all the functions of the type $c \cdot \rho(x)$, where $c > 0$ and $\rho(x) \in F$. By definition, every function from the family F is also an element of G—to show this, it is sufficient to take $c = 1$.

We will prove the desired form for all the function from the class G. This will automatically imply that all the functions from the family F also have this property.

What is the dimension of the family G, i.e., how many parameters do we need to specify each function from this family? To describe a function from the family G, we need to specify:

- the value c (1 parameter), and
- the function $\rho(x) \in F$—which requires n parameters.

Thus, $n + 1$ parameters are sufficient, and the dimension of the family G is $\leq n + 1$.
2°. For the family G, the first property of the family F—allowing combining knowledge—leads to a simpler property: that for every two functions $f_1(x)$ and $f_2(x)$ from the family G their product $f_1(x) \cdot f_2(x)$ also belong to G.

Indeed, the fact that each function $f_i(x)$ belongs to G means that it has the form $c_i \cdot \rho_i(x)$ for some $c_i > 0$ and for some function $\rho_i(x)$ from the family F. Thus, the product $f(x) = f_1(x) \cdot f_2(x)$ of these functions has the from $f(x) = c_1 \cdot c_2 \cdot \rho_1(x) \cdot \rho_2(x)$. By the property of allowing combining knowledge, for some $c > 0$, the function $\rho_0(x) = c \cdot \rho_1(x) \cdot \rho_2(x)$ also belongs to the family F. Thus, we have

$$f(x) = \frac{c_1 \cdot c_2}{c} \cdot (c \cdot \rho_1(x) \cdot \rho_2(x)) = c_0 \cdot \rho_0(x),$$

where we denoted $c_0 \overset{\text{def}}{=} \dfrac{c_1 \cdot c_2}{c}$. So indeed, $f(x) \in G$.
3°. Similarly, from the other properties of the family F, we can make the following conclusions:

- that for every function $f(x)$ from the family G and for every natural number n, the function $(f(x))^{1/n}$ also belongs to G;
- that for every function $f(x)$ from the family G and for every real number a, the function $f(x + a)$ also belongs to G; and
- that for every function $f(x)$ from the family G and for every real number $\lambda > 0$, the function $f(\lambda \cdot x)$ also belongs to G.

6°. We can simplify the problem even more if instead of the family G, we consider the family g of all the functions of the type $F(x) = \ln(f(x))$, where $f(x) \in G$. To such functions, we also add the limit functions.

Adding limit cases does not increase the dimension, so the dimension of the family g is still $\leq n + 1$.

In terms of this new family, we need to prove that all the functions from this family are polynomials of order $\leq n$.

The fact that the family G is closed under multiplication means that the family g is closed under addition. The fact that the family G is closed under taking the n-th root means that the family g is closed under multiplication by $1/n$ for each natural number n. Together with closing under addition, this means that for every two natural numbers m and n, the function

$$\frac{m}{n} \cdot F(x) = \frac{1}{n} \cdot F(x) + \cdots + \frac{1}{n} \cdot F(x) \ (m \text{ times})$$

also belongs to the family g. In other words, for every function $F(x) \in g$ and for every rational number r, the product $r \cdot F(x)$ also belongs to g. Since every real number is a limit of rational numbers—e.g., of numbers obtained if we only keep the first N digits in the decimal or binary expansion—and we added all limit cases, we can conclude that $r \cdot F(x) \in g$ for all non-negative real numbers r as well.

One can easily show that shift- and scale-invariance properties are also satisfied for the new family:

- that for every function $F(x)$ from the family g and for every real number a, the function $F(x + a)$ also belongs to g; and
- that for every function $F(x)$ from the family G and for every real number $\lambda > 0$, the function $F(\lambda \cdot x)$ also belongs to g.

$7°$. As a final simplification, we consider the family h of all the differences $d(x) = F_1(x) - F_2(x)$ between functions from the class g. To describe each of the functions $F_1(x)$ and $F_2(x)$, we need $n + 1$ parameters, so the dimension of the new family does not exceed $2 \cdot (n + 1)$.

Since for every function $F(x) \in g$, the function $2F(x)$ also belongs to the family g, we can conclude that the difference $F(x) = (2F(x)) - F(x)$ also belongs to the family h. Thus, $g \subseteq h$.

The family h is also closed under addition. Indeed, if $d_1(x) = F_{11}(x) - F_{12}(x)$ and $d_2(x) = F_{21}(x) - F_{22}(x)$ for some $F_{ij}(x) \in g$, then

$$d_1(x) + d_2(x) = (F_{11}(x) - F_{12}(x)) + (F_{21}(x) - F_{22}(x)) =$$

$$(F_{11}(x) + F_{21}(x)) - (F_{12}(x) + F_{22}(x)),$$

where, since g is closed under addition, the sums $F_{11}(x) + F_{21}(x)$ and $F_{12}(x) + F_{22}(x)$ also belong to g. Thus, indeed, the sum $d_1(x) + d_2(x)$ is a difference between two functions from g and is, thus, an element of the family h.

We can also prove that the family h is closed under multiplication by any real number c. Indeed, let $d(x) = F_1(x) - F_2(x)$.

- If $c > 0$, then $c \cdot d(x) = (c \cdot F_1(x)) - (c \cdot F_2(x))$, where both $c \cdot F_1(x)$ and $c \cdot F_2(x)$ belong to the family g.
- If $c < 0$, then $c \cdot F(x) = |c| \cdot F_2(x) - |c| \cdot F_1(x)$, where also $|c| \cdot F_2(x)$ and $|c| \cdot F_1(x)$ belong to the family g.

So, the family g is closed under addition and under multiplication by any real number—and is, thus, a linear space. Let $d \leq 2n + 2$ denote the dimension of this linear space, and let us select a basis $e_1(x), \ldots, e_d(x)$. This means that all functions from the space g have the form $c_1 \cdot e_1(x) + \cdots + c_d \cdot e_d(x)$.

From the fact that the family g is shift- and scale-invariant, we can conclude that the family h is also shift- and scale-invariant.

$8°$. As we have shown in Chap. 3, the fact that the linear space is shift- and scale-invariant implies that all the functions from this space h are polynomials. Since $g \subseteq h$, all the functions from the class g are also polynomials.

9°. What is the order of these polynomials? Let D be the order of a polynomial $F(x)$ from the class g. For each polynomial of order D, in general, the functions $F(x)$, $F(x+h)$, $F(x+2h)$, ..., $F(x+D \cdot h)$ are linearly independent: indeed, for $h \to 0$, this is equivalent to linear independence of x^D, x^{D-1}, ..., 1, and thus, in the generic case, the corresponding determinant is different from 0. Since we have $D+1$ independent functions, thus, the family g has dimension $D+1$. But we know that the dimension of this family is $\leq n+1$. From $D+1 \leq n+1$, we conclude that $D \leq n$. Thus, all functions $F(x) = \ln(f(x))$ from the class g are polynomials of order $\leq n$. Hence, each function $f(x) = \exp(F(x))$ from the class F has the desired form.

The proposition is proven.

35.3 Consequence

Empirical distributions: we expect them to be multi-modal. Continuous distributions are characterized by their probability density functions $\rho(x)$. In principle, a probability density function can be any non-negative function, the only condition is that the overall probability should be equal to 1, i.e., that

$$\int \rho(x)\,dx = 1.$$

In such situations, it is natural to expect that, in general, we will observe generic functions with this property—e.g., functions which are random with respect to some reasonable measure on the set of all functions. The first such measure was Wiener measure, corresponding to random walk. Later, many other random measures have been proposed. In most of these random measures, almost all functions are truly random, similar to random walk—in the sense that are very "wiggly", they have infinitely many local maxima and minima. In probabilistic terms, we expect the empirical probability density functions to be multi-modal.

Empirical distributions are mostly few-modal. In reality, empirical distributions are mostly either unimodal, or bimodal, or—in rare cases—trimodal. In other words, they are usually few-modal; see, e.g., [2]. Why?

This is especially puzzling in econometrics. In science and engineering, the few-modality is often easy to explain: e.g., the distributions are normal or Gamma, or, in general, follow some theoretically justified law. But few-modal distributions are ubiquitous also in situations where we do not have exact equations—such as econometrics. Why?

What we do. The above result provides an explanation for this empirical fact.

Corollary. *Every function from a shift- and scale-invariant n-parametric family of distributions that allows combining knowledge and partial knowledge has no more than n − 1 local maxima and local minima.*

Proof of the Corollary. Indeed, at local maxima and minima, the derivative $\rho'(x) = \exp(P(x)) \cdot P'(x)$ is equal to 0, which is equivaent to $P'(x) = 0$. The derivative $P'(x)$ is a polynomial of degree $\leq n - 1$, and such polynomials can have no more than $n - 1$ zeros.

Discussion. This explain why empirical distribution are few-modal.

References

1. Urenda, J.C., Kosheleva, O., Kreinovich, V.: Why most empirical distributions are few-modal. In: Ngoc Thach, N., Thanh Ha, D., Duc Trung, N., Kreinovich, V. (eds.), Prediction and Causality in Econometrics and Related Topics
2. Sheskin, D.J.: Handbook of Parametric and Non-Parametric Statistical Procedures. Chapman & Hall/CRC, London, UK (2011)

Chapter 36
Data Processing: Probabilistic Techniques, Part 3

In the previous two chapters, we discussed how algebraic approach can help apply probabilistic techniques. This does not mean that all the related challenges are over. In this chapter, we provide an example of such a challenge.

We expect that the quality of experts' decisions increases with their experience. This is indeed true for reasonably routine situations. However, surprisingly, empirical data shows that in unusual situations, novice experts make much better decisions than more experienced ones. This phenomenon is especially unexpected for medical emergency situations: it turns out that the mortality rate of patients treated by novice doctors is a third lower than for patients treated by experience doctors. In this chapter, we provide a possible explanation for this seemingly counterintuitive phenomenon— namely, we show this phenomenon naturally appears if we use standard probabilistic techniques.

Results from this chapter first appeared in [1]

36.1 Problem: Experts Don't Perform Well in Unusual Situations

At first glance, it would seem that the more experienced the experts, the better their solutions. This is indeed the case for reasonably routine situations. However, somewhat surprisingly, empirical data shows that in unusual situations, novice experts perform, on average, better than more experienced once. This is especially surprising in cases of medical emergency: in cases when high-risk patients were treated by relative novices, the mortality rate was a third lower than when patients were treated by experienced professional; see, e.g., [2, 3].

In this chapter, we provide a possible explanation for this counterintuitive phenomenon.

J. C. Urenda and V. Kreinovich, *Algebraic Approach to Data Processing*, Studies in Big Data 115, https://doi.org/10.1007/978-3-031-16780-5_36

36.2 Our Explanation

Formulation of the problem in precise terms. We have an observed value $\widetilde{x}$ of the desired quantity x. Based on this observation and on the expert's knowledge, we need to estimate the actual value x.

The observed value is imprecise, due to measurement and observation errors: the difference $\widetilde{x} - x$ is, in general, different from 0. Based on the prior experience, we can estimate the mean and the standard deviation of this difference. If the corresponding mean is different from 0, this simply means that our measuring instrument has a bias. In this case, we can re-calibrate this instrument, by subtracting this bias from all the measurement results. Thus, without losing generality, we can safely assume that the bias has been eliminated, so the mean is 0. Let us denote the corresponding standard deviation by σ.

The expert's experience-based knowledge consists of several cases for which this expert knows a reasonably accurate value of the corresponding quantity. For example, in the medical situation, an experienced expert is familiar with many cases in which, later on, the value of the corresponding quantity was measured more accurately. Let us denote the corresponding values by $\widetilde{x}_1, \ldots, \widetilde{x}_n$. These values are very accurate, in the sense that they are very close to the corresponding actual values x_i. However, all situations are different. As a result, the actual values x_i are, in general, different from x, and thus, the measurement results $\widetilde{x}_i$ provide only an approximate estimation for the desired quantity x. Again, from the previous experience, we can estimate the standard deviation σ_0 of the corresponding difference $x_i - x$.

Thus, we arrive at the following problem:

- we know the value $\widetilde{x}$ which is approximately equal to x with standard deviation σ:

$$\widetilde{x} \approx x \text{ (st.dev. } \sigma);$$

and
- we know n values $\widetilde{x}_1, \ldots, \widetilde{x}_n$ which are close to x with standard deviation σ_0:

$$\widetilde{x}_i \approx x \text{ (st.dev. } \sigma_0).$$

Based on these values, we must provide an estimate for the desired quantity x.

Let us solve this problem. A natural way to solve the above problem if to use the Least Squares technique (see, e.g., [4], i.e., to minimize the expression

$$\frac{(\widetilde{x} - x)^2}{\sigma^2} + \sum_{i=1}^{n} \frac{(\widetilde{x}_i - x)^2}{\sigma_0^2}.$$

Differentiating this expression by x and equating the derivative to 0, we conclude that for the resulting estimate $\widehat{x}$, we have

$$\frac{2 \cdot (\widehat{x} - \widetilde{x})^2}{\sigma^2} + \sum_{i=1}^{n} \frac{2 \cdot (\widehat{x} - \widetilde{x}_i)^2}{\sigma_0^2} = 0.$$

Multiplying both sides of this equation by $\sigma^2/2$ and moving all the terms not containing $\widehat{x}$ to the right-hand side, we conclude that

$$\widehat{x} \cdot (1 + k \cdot n) = \widetilde{x} + k \cdot \sum_{i=1}^{n} \widetilde{x}_i,$$

where we denoted $k \stackrel{\text{def}}{=} \dfrac{\sigma_0^2}{\sigma^2}$. Thus,

$$\widehat{x} = \frac{\widetilde{x} + k \cdot \sum\limits_{i=1}^{n} \widetilde{x}_i}{1 + k \cdot n}. \tag{36.1}$$

From the expression (36.1), we see that for this estimate, we do not need to know the individual values $\widetilde{x}_i$, it is sufficient to know their sum—i.e., equivalently, to know their arithmetic average

$$\overline{x} = \frac{1}{n} \cdot \sum_{i=1}^{n} \widetilde{x}_i.$$

In terms of this average, the sum $\sum\limits_{i=1}^{n} \widetilde{x}_i$ takes the form $n \cdot \overline{x}$ and thus, the formula (36.1) takes the form

$$\widehat{x} = \frac{\widetilde{x} + k \cdot n \cdot \overline{x}}{1 + k \cdot n}. \tag{36.2}$$

How accurate is the resulting estimate? To analyze how accurate is this estimate, we need to consider the absolute value Δ of the difference between this estimate and the actual value x. Here,

$$\Delta = \left| \frac{\widetilde{x} + k \cdot n \cdot \overline{x}}{1 + k \cdot n} - x \right|.$$

Explicitly subtracting the fractions in the right-hand side of this formula, we get the formula

$$\Delta = \left| \frac{\widetilde{x} + k \cdot n \cdot \overline{x} - x + k \cdot n \cdot x}{1 + k \cdot n} \right|,$$

i.e., regrouping the terms in the numerator, the formula

$$\Delta = \left| \frac{\widetilde{x} - x}{1 + k \cdot n} + \frac{k \cdot n \cdot (\overline{x} - x)}{1 + k \cdot n} \right|. \tag{36.3}$$

Let us show that this formula enables us to explain the empirical phenomenon: namely, that while experienced experts make better decisions in routine situations, their decisions in unusual situations are worse than the decisions of novice experts.

What happens in reasonably routine situations. In reasonably routine situations, when x is close to the average $\overline{x}$ and thus, when the difference $\overline{x} - x$ is very small, the first term in the right-hand side of the formula (36.3) dominates, so we have

$$\Delta \approx \frac{|\widetilde{x} - x|}{1 + k \cdot n}.$$

In this case, the more experienced the expert, i.e., the larger the corresponding value n, the smaller the value Δ and thus, the more accurate the expert's estimate. So, in such cases, indeed, the more experienced the expert, the more accurate his/her estimates.

What happens in unusual situations. Let us now consider unusual situations, when the difference $\overline{x} - x$ is large, so that the absolute value $|\overline{x} - x|$ of this difference is larger than the typical observation uncertainty $|\widetilde{x} - x| \approx \sigma$:

$$|\overline{x} - x| > |\widetilde{x} - x|. \tag{36.4}$$

In this case, for experienced experts, for which n is large, the second term in the formula (36.3) dominates, so we get

$$\Delta_{\exp} \approx |\overline{x} - x|. \tag{36.5}$$

In contrast, for novice experts, e.g., for experts with $n = 0$ (e.g., medical doctors who have just received their degrees and do not have the experience of independently treating patients), we have

$$\Delta_{\text{nov}} = |\widetilde{x} - x|. \tag{36.6}$$

Comparing the expressions (36.5) and (36.6) and taking into account the inequality (36.4), we conclude that here, indeed, $\Delta_{\text{nov}} < \Delta_{\exp}$, i.e., that in such unusual situations, novice experts indeed make more accurate estimates (and thus, better decisions) that experienced ones.

References

1. Urenda, J.C., Kreinovich, V.: Why experts sometimes do not perform well in unusual situations. Math. Struct. Model. **51**, 109–113 (2019)
2. Jena, A.B., Prasad, V., Goldman, D.P., Romley, J.: Mortality and treatment patterns among patients hospitalized with acute cardiovascular conditions during dates of national cardiology meetings. J. Amer. Med. Assoc. JAMA: Internal Med. **175**(2), 237–244 (2015)

3. Mlodinow, L.: Elastic: Flexible Thinking in a Time of Change. Pantheon Books, New York (2018)
4. Sheskin, D.J.: Handbook of Parametric and Non-Parametric Statistical Procedures. Chapman & Hall/CRC, London (2011)

Chapter 37
Data Processing: Beyond Traditional Techniques

The idea of neural networks comes from simulating human brains—which are often very successful in solving problems. But the same example of a human brain shows that there are other successful processes worth emulating: namely, neural networks emulate how brain *works*, but other aspects are related to how the brain emerges. On the molecular level, all the cells come from DNA-related activities. Simulating such activity is the essence of *DNA computing*.

The traditional DNA computing schemes are based on using or simulating DNA-related activity. This is similar to how quantum computers use quantum activities to perform computations. Interestingly, in quantum computing, there is another phenomenon known as *computing without computing*, when, somewhat surprisingly, the result of the computation appears without invoking the actual quantum processes. In this chapter, we show that similar phenomenon is possible for DNA computing: in addition to the more traditional way of using or simulating DNA *activity*, we can also use DNA *inactivity* to solve complex problems. We also show that while DNA computing without computing is *almost as powerful* as traditional DNA computing, it is actually *somewhat less powerful*. As a side effect of this result, we also show that, in general, security is somewhat more difficult to maintain than privacy, and data storage is more difficult than data transmission.

Results from this chapter first appeared in [1].

37.1 DNA Computing: Introduction

In his famous 1994 paper [2], Leonard Adleman showed that, in principle, we can drastically speed up computations if we use the fact that DNA fragments combine together—in a process known as *ligation*—if we corresponding nucleotides match, i.e., if:

J. C. Urenda and V. Kreinovich, *Algebraic Approach to Data Processing*,
Studies in Big Data 115, https://doi.org/10.1007/978-3-031-16780-5_37

- A is matched with T,
- T is matched with A,
- C is matched with G, and
- G is matched with C.

For example, two fragments ACTTG and TGAAC match perfectly.

Specifically, this paper showed that we can speed up the solution to the following *Hamiltonian path problem*:

- given a graph,
- find a path in this graph that visits every vertex exactly once.

This seminal paper started the field of *DNA computing*, which now includes both:

- using actual DNA fragments (as Adleman did) and
- using computer simulation of the corresponding processes.

One of the main advantages of computing via molecular interactions, when each molecule serves as a processor, is that in each mole, we have 10^{23} molecules— and thus, 10^{23} processor working in parallel. Such unbelievable parallelism— many orders of magnitude higher than the usual thousands of processors in a supercomputer—is a clear indication that this approach has a great potential.

Later, similar DNA-based algorithms were proposed for solving other complex problems, such as propositional satisfiability (this problem is explained, in detail, later in this chapter). For reasonably recent overviews, see, e.g., [3–7].

All these algorithms are based on actually using (or simulating) the ligation process. This is similar to how quantum computers use quantum activities to perform computations. Interestingly, in quantum computing, there is another phenomenon known as *computing without computing*, when, somewhat surprisingly, the result of the computation appears without actually invoking quantum processes. In this chapter, we show that similar phenomenon is possible for DNA computing:

- in addition to the more traditional way of using or simulating DNA *activity*,
- we can also use DNA *inactivity* to solve complex problems.

37.2 Computing Without Computing—Quantum Version: A Brief Reminder

DNA computing is one of several directions in the general quest for using novel physical phenomena in computing. Another—probably even more well known— direction is *quantum computing*, the use of quantum effects to speed up computations; see, e.g., [8, 9].

Most quantum algorithms actually use quantum effects to perform the corresponding computations, but there is an interesting version called *counterfactual quantum computing*, or, alternatively, *computing without computing*. The idea is that:

- we *set up* the corresponding quantum computations, but
- we *do not* actually *run* them,

and still, because of the quantum effects, we get the desired result with some probability.

This idea was first proposed in [10]. The main motivation behind this idea was not so much about *computing* but rather about *testing*: the same idea can be, in principle, used to test the complex equipment without actually running it. For example, in principle, we can test whether the atomic bomb (that has been in storage for a long time) will actually explode when triggered—without actually having to explode it to find this out.

At this moment, this quantum computing-without-computing phenomenon is far from practical use—just like most quantum computing algorithms and most DNA computing algorithms are still far from practical use. However, there has been a lot of progress in this direction. For example:

- Initially, there was a fear that the probability of getting the correct result in the computing-without-computing setting may be too low to be practically useful.
- However, in 2006, a seminal paper [11] showed that this probability can be increased to almost 1.

The fact that in quantum computing, it is possible to perform some computations without actually running these computations encouraged us to check whether a similar phenomenon is possible for DNA computing as well. We were even further encouraged by the fact that computing without computing is also theoretically possible in yet another direction of using novel physical phenomena in computing—namely, in the use of acausal effects. Let us briefly recall this idea.

37.3 Computing Without Computing—Version Involving Acausal Processes: A Reminder

How can we speed up computations? A natural science-fiction idea is to use a *time machine* (also known as an *acausal*—i.e., causality violating—process):

- we let the computer spend as much time as needed, even it means several thousand years, and then
- we use the time machine to bring these results back to us.

For a long time, acausal processes remained mostly the subject of science fiction. Serious physicists mostly believed that time machines are not possible—due to well known paradoxes. These paradoxes can be summarized by stating the probably well known paradox of time travel—the grandfather paradox: what if a time traveler goes into the past and kills his own grandfather before the traveler's parents are conceived?

In spite of the paradoxes, acausal processes continued to naturally emerge in many areas of physics. This emergence is mostly related to the fact that:

- in contrast to pre-quantum physics, where everything is deterministic,
- in quantum physics, we can only make probabilistic predictions.

In other words, there are always fluctuations, deviations of the actual values from the expected values of the corresponding physical quantities.

In pre-quantum physics, at each moment of time, a particle is in a certain spatial location, with a certain velocity—and, in principle, we can measure both location and velocity with any desired accuracy. In quantum physics, such exact measurements are no longer possible. A particle's location and velocity are always probabilistic: e.g., even if we prepare several particles in the identical states and measure their velocities, we will get slightly different results. And the smaller the region we consider, the larger these fluctuations.

Similarly, the space-time tensor—that describes the geometry of space-time and the direction of causality—fluctuates. The smaller the region we consider, the larger these fluctuations. As a result, the maximal possible speed fluctuates from the usual macroscopic speed-of-light value c:

- in some microscopic locations, the maximal speed is slightly larger than c, while
- in some other microscopic locations, the maximal speed is slightly smaller than c.

If a micro-particle follows the locations when the local maximal speed is larger than c, then, from the macroscopic viewpoint, this perfectly physical particle goes faster than the speed of light—and, according to special relativity, this implies the possible of going back in time.

Many other schemes naturally appeared in physics, thus leading to acausal effects. As a result, in the late 1980s, a group of physicists led by a future Nobelist Kip Thorne decided to overcome the previous taboo and to seriously analyze possible acausal processes; see, e.g., [12–15].

But what about the paradoxes? Here, the probabilistic nature of quantum physics also helps. As we have mentioned, in quantum world, nothing is guaranteed. If the time traveler attempted to kill his grandfather; then:

- since the grandfather was alive enough after that attempt to sire a son,
- this means that this attempt failed.

In other words, some event happened which prevented the killing:

- maybe a policeman walked by and prevented the murder,
- maybe the gun got stuck,
- maybe a meteorite fell on the gun at that exact moment.

We can try to prevent all such events, but no matter how much we try, no matter how many possibilities we take into account, there is always a possibility of some rare, low-probability event that would disrupt the process. So, the only real consequence of trying to implement a time-travel paradox is that some very low-probability event will happen.

And, interestingly, this can be used to computations—i.e., we can use the *possibility* of acausal effects to perform computations without actually invoking these effects. In other words, we have another case of computing without computing. Indeed, suppose, e.g., that:

- we are given a graph, and
- we need to find a Hamiltonian path in this graph.

What we can do is:

- use a random number generators to generate some (random) path through this graph, and then
- check if the resulting path is Hamiltonian.

If the path is not Hamiltonian, we launch a time machine—which is set up in such a way that its launch leads to some low-probability event, with probability $p_0 \ll 1$.

On the other hand, e.g., in a binary graph, the probability that a random selection of a direction at each of n nodes will lead to a selected path is 2^{-n}. So, nature has a choice:

- it can set up random processes so as to select a Hamiltonian path, or
- it will have to implement a low-probability event, with probability $p_0 \ll 1$.

According to the general idea of statistical physics, in most cases, nature selects the event with higher probability. So, if $p_0 \ll 2^{-n}$, nature will select a Hamiltonian path—and thus, we will find this path fast without actually having to use the time machine.

Comments.

- This idea is described, e.g., in [16–23].
- Now that we have learned how computing without computing is possible in quantum and acausal computing, let us show how (and why) this idea is possible in DNA computing as well.

37.4 Computing Without Computing—DNA Version

Main idea. Let us show that with DNA computing, it is also possible to solve complex problems by using or simulating DNA inactivity.

The possibility of inactivity makes perfect biological sense:

- when resources are plentiful, it makes sense for the living creatures to be active and to actively multiply, but
- in situations when resources become scarce, such an activity would exhaust these resources really fast.

In such situations, it is important to slow down all the biological processes as much as possible.

In nature, we observe such slowing down all the time:

- from hibernating bears

- to plants that stop practically all activities in winter
- to bacteria and viruses that can slow down to such an extent that they can survive in this slowed-down condition for hundreds and even thousands of years.

The slow-down occurs on all the levels:

- from the macro level—when an animal (e.g., a hibernating bear) stops moving almost completely,
- to the cell level, where all the usual biochemical processes grind practically to a halt.

On the DNA level, this means that instead of enhancing the possible ligations, in such situations, the cell tries to prevent ligation as much as possible, so as to keep all the processes inactive. This phenomenon has indeed been traced on the gene expression level; see, e.g., [24]. The possibility for such prevention comes from the fact that:

- contrary to a somewhat simplified version of DNA processes used in the traditional DNA computing,
- the actual DNA-related biochemical processes do not simply involve matching of different parts of the RNA and DNA.

There is also a *control* that switches some genes (i.e., some parts of the RNA and DNA) on and off. This control is determined:

- partly by other genes, and
- partly by the signals that the cell gets from the environment.

From this viewpoint, in the case of scarce resources, the corresponding control processes are organized in a way to maximally prevent ligations.

We will describe this control process in precise terms, and let us show that the corresponding problem is NP-hard—which means that it can be used to solve complex computational problems. But before we do that, let us explain why we believe that such control can be used for computations.

It is not easy to stop biological processes. The great potential of DNA computing comes from the fact that the corresponding biological processes are very complicated. In spite of the original optimism, even though the genomes of many living creatures—including humans—have been decoded, we are almost as far from the full understanding the corresponding processes as before—and even farther from artificially synthesizing even the simplest living creatures. The problems are complex, but within each of numerous cells of numerous living creatures, nature solves the corresponding complex problems all the time. Thus, it is natural to try to use these naturally occurring solutions to solve our complex problems.

DNA processes are complex, but nature knows how to solve them—and thus, they occur all the time. Stopping these processes is much more difficult, even for nature—indeed, very few living creatures can do it, and we are still far from understanding how this is done.

- A grain left outside eventually spoils and rots, but some grains got preserved for thousands of years—and when planted, turned into plants.
- Freezing kills most living creatures, but some mysteriously survive—and get revised when thawed out.
- Viruses and bacteria can survive for years in the cosmic cold—there is even a *panspermia* hypothesis that this is how life spreads between the planets, this is how originated on Earth.
- The possibility to stop biological processes in a human being—known as *anabiosis*—is a common feature in science fiction, but in real life, it remains a far-from-possible dream.

Since stopping of biological processes is too difficult, even more difficult that running them, it is even more reasonable to use this stopping—in addition to the DNA processes themselves—to solve other complex problems.

Towards describing ligation prevention in precise terms. In general, we have several fragments that, in principle, have matching parts. Each fragment consists of several sub-fragments, and we can decide which of these sub-fragments is switched on to be active. We want to select the sub-fragments in such a way that no two active sub-fragments are matched.

Here is a precise formulation of the problem.

What is given. We have several (N) nucleotide sequences ("fragments") s_1, ..., s_N, i.e., sequences consisting of symbols C, G, A, and T. Each fragment s_i is a concatenation of several subsequences ("sub-fragments") $s_i = s_{i1} \ldots s_{ik_i}$.

The sub-fragments s and s' *match* (or are *complementary*) if s' can be obtained from s by replacing A with T, T with A, C with G, and G with C.

What we want to find. The problem is to find the integers $j_1, \ldots, j_N$ such that $1 \leq j_i \leq k_i$ and for every two fragments i and i', the corresponding sub-fragments s_{ij_i} and $s_{i'j_{i'}}$ do not match.

Let us prove that the ligation prevention problem is NP-hard. In practice, we are usually interested in the problems in which, once someone provides us with a candidate for a solution, we can feasibly tell whether this is a solution or not. The class of all such problems is known as the class NP; see, e.g., [19, 25].

Some computational problems are NP-hard, meaning that every problem from the class NP can be reduced to this problem. In other words, if we have an efficient algorithm for solving an NP-hard problem, this means that by reducing to this problem, we can solve *any* practical problem in feasible time [19, 25].

If a problem is NP-hard *and* itself belongs to the class NP, then this general problem is known as *NP-complete*.

Let us show that the ligation prevention problem is NP-hard. Since it is easy to check that no two sub-fragments are complementary to each other, this means that this problem is also in the class NP and is, thus, actually NP-complete.

This would mean that, if—as we believe—nature has a way to solve the ligation prevention problem (at least many instances of this problem), then by reducing to this problem, we will be able to solve many practical problems in reasonable time.

How NP-hardness is usually proved. To show that a given problem P_{given} is NP-hard, it is sufficient to show that a known NP-hard problem P_{known} can be reduced to this problem. Indeed, by definition of NP-hardness, every problem P from the class NP can be reduced to P_{known}, and since the problem P_{known} can be, in its turn, reduced to P_{given}, this would mean that a two-stage reduction $P \rightarrow P_{known} \rightarrow P_{given}$ reduces P to P_{given}. Since this is true for every problem P from the class NP, this means that the given problem P_{given} is indeed NP-hard.

How we will prove NP-hardness. As the known problem P_{known}, we select the propositional satisfiability problem for 3-CNF formulas, historically the first problem proven to be NP-hard. In this general problem, we deal with *Boolean* (= *propositional*) variables, i.e., variables $x_1, \ldots, x_v$ that can take two possible values: 1 (meaning "true") and 0 (meaning "false"). A *literal a* is either a variable x_k or its negation $\neg x_k$.

A *clause* is an expression of the type $a \lor b$ or $a \lor b \lor c$ where a, b, and c are literals. Examples are $x_1 \lor \neg x_2$ or $\neg x_1 \lor \neg x_5 \lor x_9$.

Finally, a *propositional formula F* (or simply a *formula*, for short) is an expression of the type $C_1 \& C_2 \& \ldots \& C_m$, where C_i are clauses. An example of a formula is the expression

$$(x_1 \lor \neg x_2) \& (\neg x_1 \lor \neg x_5 \lor x_9).$$

The general propositional satisfiability problem is:

- given a formula,
- find the values of the variables that make it true (or, to be more precise, to check whether there exist values x_i that make it true).

The actual proof by reduction. Let us assume that we are given an instance F of the propositional satisfiability problem, i.e., that we are given a propositional formula F of the type $C_1 \& \ldots \& C_m$ with v boolean variables $x_1, \ldots, x_v$.

To reduce this instance to an appropriate instance of the ligation prevention problem, first, we assign, to each boolean variable x_j, a fragment $f(x_j)$ consisting of letters C, G, A, and T, in such a way that fragments assigned to two different variables are not complementary.

There are many ways to do it. For example, we can assign v different fragments to v variables, and then add a letter A in front of each of these fragments. This way, no two fragments will fully match, since for them to match, their first symbols must match as well, but A does not match with A—it only matches with T.

To each negation $\neg x_j$, we assign a fragment—which we will denote by $f(\neg x_j)$—which is complementary to $f(x_j)$, i.e., which is obtained from $f(x_j)$ by replacing A with T, T with A, C with G, and G with C.

Finally, to each clause C_i, we assign a fragment s_i in the following way:

- if the clause has the form $a \lor b$, then we take a fragment $s_i = f(a)f(b)$ consisting of two sub-fragments $f(a)$ and $f(b)$;
- if the clause has the form $a \lor b \lor c$, then we take a fragment $s_i = f(a)f(b)f(c)$ consisting of three sub-fragments $f(a)$, $f(b)$, and $f(c)$.

Let us show that the original formula F is satisfiable if and only if it is possible to select a sub-fragment in each fragment s_i so that none of the selected sub-fragments are complementary to each other.

Indeed, if the formula F is satisfiable, this means that there exists an assignment of truth values to all the boolean variables $x_1, \ldots, x_v$ that makes the formula F true—which means that each of the clauses C_i is true. The fact that a clause C_i is true means that one of its literals is true. We thus select a sub-fragment corresponding to one of the true literals.

No two selected sub-fragments are complementary to each other—indeed, complementary would mean that they represent a variable x_j and its negation $\neg x_j$, and the variable and its negation cannot be both true.

Vice versa, let us assume that we for each fragment s_i corresponding to a clause $C_i = a \vee \ldots$, we selected a sub-fragment—let us denote it by $f(a_i)$—so that no two sub-fragments are complementary to each other. The fact that they are not complementary means that no two corresponding literals a_i and $a_{j'}$ are negations of each other. Thus, we can assign the truth value to each of the boolean variables x_j as follows:

- if one of the selected sub-fragments has the form $f(x_j)$, then we make the boolean variable x_j true;
- if one of the selected sub-fragments has the form $f(\neg x_j)$, then we make the boolean variable x_j false;
- if none of the selected sub-fragments is of the form $f(x_j)$ or $f(\neg x_j)$, then we assign any truth value to x_j.

Since no two selected sub-fragments have the form $f(x_j)$ and $f(\neg x_j)$, this means that this assignment is consistent. In this assignment, for each clause C_i, the literal a_i corresponding to the selected sub-fragment $f(a_i)$ is true. Thus, under this assignment, each clause C_i is true and hence, the whole formula $F = C_1 \& \ldots \& C_m$ is true.

The reduction is proven.

Comment. While we reduced propositional satisfiability to our problem, in fact, this proof can be viewed as reducing another NP-complete problem to our problem—namely, the problem of finding a *clique* of given size k in a given graph. A clique is defined as a subset of the graph's vertices in which every two vertices are connected to each other by an edge. Our proof is actually a modification of the standard proof that the clique problem is NP-complete; see, e.g., [25].

In this proof, we reduce the propositional satisfiability problem to the clique problem in the following way. Let an instance F of the propositional satisfiability problem be given. This instance has the form $C_1 \& \ldots \& C_m$, where C_i are clauses. For each literals a from each clause C_i, we add a vertex $V_i(a)$ to the resulting graph.

For example, for the formula $(x_1 \vee \neg x_2) \& (x_1 \vee x_2 \vee x_3)$, we have a graph with five vertices:

- two vertices $V_1(x_1)$ and $V_1(\neg x_2)$ corresponding to the first clause, and
- three vertices $V_2(x_1)$, $V_2(x_2)$, and $V_2(x_3)$ corresponding to the second clause.

We then connect, by edges, vertices corresponding to different literals provided that they do not correspond to opposite literals x_i and $\neg x_i$. In the above example,

- the vertex $V_1(x_1)$ is connected to $V_2(x_1)$, $V_2(x_2)$, and $V_2(x_3)$; and
- the vertex $V_1(\neg x_2)$ is connected to $V_2(x_1)$ and $V_2(x_3)$ (but *not* to $V_2(x_2)$).

The fact that this is indeed a reduction can be easily proven.

Indeed, if the original formula F is satisfiable, then in each clause, (at least) one of the literals is true. We can select one true literal in each clause. These literals cannot be opposite: since we cannot have x_i and $\neg x_i$ both true. Thus, every two corresponding vertices are connected—i.e., the resulting subgraph indeed forms a clique of size m.

Vice versa, if we have a clique of size m, then, since literals corresponding to the same clause are not connected, this means that vertices from this clique correspond to different clauses. And since we have exactly m clauses, this means that the clique contains exactly one vertex corresponding to each clause. Now, we can select, for each variable x_i, the value "true" or "false" depending on whether the clique contains a vertex corresponding to x_i or a vertex corresponding to $\neg x_i$. The clique cannot contain both—since vertices corresponding to opposite literals are not connected. (For the variables not reflected in any of the vertices from the clique, we can select any truth value.)

Since each clause C_i contains at least one vertex $V_i(a)$ from the clique, the corresponding literal a is true in this assignment, and thus, the clause C_i is also true. So, under this assignment, all clauses are true—and hence, the original formula $F = C_1 \& \ldots \& C_m$ is also true.

The reduction is proven.

37.5   DNA Computing Without Computing Is Somewhat Less Powerful Than Traditional DNA Computing: A Proof

Which of the two DNA computing schemes is more powerful? In the previous section, we have shown that, in addition to the traditional DNA computing that utilizes the actual DNA-related chemical processes, we can also perform effective computations by using the ability of a body to stop these chemical processes. A natural question is: which of the two DNA computing schemes is more powerful: the active or the passive one?

Overall, they are both NP-complete, in this sense they are both equally powerful. However, we can still talk about which problems are more powerful and which problems are less powerful if we take into account a subtle subdivision of NP-complete problems.

W-hierarchy: a brief reminder. The subtle subdivision that we have in mind—called *W-hierarchy*—is based on the notions of *fixed parameter tractable* (fpt) problems and of *weft*. We will briefly explain these notions in this section; readers interested in detail can check, e.g., [26–28].

The main idea is that while a problem may be, in general, NP-hard—which means that unless it turns out that $P = NP$, we cannot have a feasible (polynomial-time) algorithm for solving this problem, there usually is a parameter k such that if we bound the value of this parameter, the problem can be solved in polynomial time, i.e., if, some computable functions $f(k)$ and $C(k)$, a problem with input x of size $n \overset{\text{def}}{=} \text{len}(x)$ can be solved in time $f(k) \cdot n^{C(k)}$. This way, if we fix some bound k_0 and only consider problems for which the value of k is bounded by k_0, then all thus limited problems can be solved in time $\leq f_0 \cdot n^{C_0}$, where $f_0 \overset{\text{def}}{=} \max(f(1), f(2), \ldots, f(k_0))$ and $C_0 \overset{\text{def}}{=} \max(C(1), C(2), \ldots, C(k_0))$.

For some problems, the corresponding exponent $C(k)$ does not grow with k. Such problems are called *fixed parameter tractable* (fpt). In precise terms, a problem is fpt if, for some computable function $f(k)$ and for some constant C, a problem with input x of size $n = \text{len}(x)$ can be solved in time $f(k) \cdot n^C$. This way, if we fix some bound k_0 and only consider problems for which the value of k is bounded by k_0, then all thus limited problems can be solved in time $\leq f_0 \cdot n^C$.

Similarly to the usual reduction, we can define *fpt-reduction* as an reduction that preserves both the size of the inputs (modulo a possible feasible—polynomial-size—increase) *and* preserves the bounds on the parameter, so that problems for which the value of the parameter is bounded by some value k_0 get transformed into problems for which the parameter is bounded by $g(k_0)$ for some feasible function $g(x)$.

The W-hierarchy is based on reduction to computational schemes of a certain *weft*. To describe the weft computation scheme, we first represent this scheme as a directed graph:

- whose vertices are elementary logical (bit) operations and
- where an edge from a vertex a to a vertex b means that the output of a is one of the inputs of the operation b.

For commutative and associative logical operations,

- in addition to the usual binary operations,
- we also allow operations with more than two inputs.

Such operations are "and", "or", and addition modulo 2 (which is the same as "exclusive or").

The weft is defined as the largest number of logical units from an input to the output. For each natural number $i = 0, 1, 2, \ldots$, the ith class $W[i]$ is defined as the class of the problems that can be reduced to a computation scheme of weft $\leq i$ with several inputs $v_1, \ldots, v_m$ and one output v for which:

- the original problem x with parameter k has a solution if and only if
- there is a combination of inputs $v_1, \ldots, v_m$ that produces the result $v =$ "true" and in which at most k inputs v_j are 1s—the rest are 0s.

It can be shown that $W[0]$ is exactly the class FPT of all fpt problems, and one can easily see that $W[0] \subseteq W[1] \subseteq W[2] \subseteq \cdots$

It is not proven that classes $W[i]$ corresponding to different i classes are indeed different, but most computer scientists believe that they *are* different, i.e., that the containment is strict: $W[0] \subset W[1] \subset W[2] \subset \cdots$ Within each class $W[i]$, there are problems which are the hardest in this class—in the sense that every other problem from this class can be fpt-reduced to this problem. Such problems are called $W[i]$-*complete*.

In particular:

- the Hamiltonian path problem—historically the first problem for which an DNA-based solution has been proposed—has been proven to be $W[2]$-complete for k being the graph width (see, e.g., [29]), while, e.g.,
- the clique problem—the problem of finding, in a given graph a clique of a given size k—is known to be $W[1]$-complete; see, e.g., [26, 27].

Since:

- the original DNA computing solves the Hamiltonian path problem while
- the DNA-based computing without computing is equivalent to the clique problem,

we thus arrive at the following conclusion.

Conclusion. The traditional DNA computing is more powerful that DNA computing without computing.

Specifically, while both traditional DNA computing and DNA-based computing without computing solve NP-complete problems:

- the traditional DNA computing is $W[2]$-complete, while
- the DNA-based computing without computing is only $W[1]$-complete, i.e., complete for the somewhat less-complex class of the W-hierarchy.

37.6   First Related Result: Security Is More Difficult to Achieve than Privacy

What we plan to do in this section. The result from the previous section can be applied to a topic which is not related to DNA computing, but which is very important: the need to maintain privacy and security when using computers.

The reason why such an application is possible is that the main problems of both privacy and security can be reformulated in graph terms.

How to describe privacy in graph terms. Privacy means that:

- while we *should* have access to our own records,
- we *should not* get unauthorized access to any other records.

This means, in particular, that:

- if we perform a simple modification of codewords and other means to get access to our own records,
- we should not be able to gain access to records of anybody else (unless that person gave us a special permission).

To describe this in graph terms, let us form a graph in which individuals are vertices, and two vertices a and b are connected if:

- it is not possible for the individual corresponding to vertex a to access b's record by a simple modification of a's access information; and,
- vice versa, it is not possible for the individual corresponding to vertex b to access a's record by a simple modification of b's access information.

Each abstract access scheme can be represented as such a general graph. The question is: can we use this abstract scheme to provide full privacy for a given number k of users? In terms of the above graph, this is equivalent to finding a subset of k vertices in which every two vertices are connected to each other—i.e., to finding a clique of the given size k.

Thus, in graph terms, maintaining privacy is equivalent to solving the clique problem. We already know that this problem is NP-complete and $W[1]$-complete.

How to describe security in graph terms. In general, computer security (and security in general) means that we have resources so that:

- if we have trouble at some location (physical or virtual),
- one of these resources is available to resolve the corresponding problem.

In the ideal world, we should have such resources at each location. However, realistically, this is usually not possible, so only some locations have resources. In terms of the police example, this means that:

- while we cannot place a police officer at every house, but
- we need to make sure that if a crime is reported, the police from the nearby police station should arrive on time to stop this crime.

Similarly, in computer security, if a suspicious message appears on a computer, the corresponding server should be able to block the corresponding virus from infecting other computers.

This situation can also be described as a graph. Namely, its vertices are possible locations. We connect the two locations a and b if these two locations are "close" in the following sense:

- a resource located at location a can reach location b in time to resolve any possible problem, and
- a resource located at location b can reach location a in time to resolve any possible problem.

Based on the geography and/or on communication ability of the corresponding network, we can form a graph of possible locations, in which edges correspond to the above "closeness". Our overall resources are limited. So, the question is:

- given that we only have k resources,
- is it possible to place them in such a way that every location is the graph is covered—i.e., that each vertex is close to one of the k selected locations?

In graph terms, the corresponding set of k locations is called a *dominating set*. In these terms, the question is: given a graph, is there a dominating set of size k in this graph? It is known that this problem is NP-complete and $W[2]$-complete; see, e.g., [30].

Conclusion: security is more difficult to maintain than privacy. Since:

- security corresponds to a $W[2]$-complete problem, and
- privacy corresponds to a $W[1]$-complete problem—which are, in general, somewhat less complex than $W[2]$-complete problems,

we can therefore conclude that security is somewhat more difficult to maintain than privacy.

37.7   Second Related Result: Data Storage Is More Difficult Than Data Transmission

Application to information science. A similar result is related to information science, the science of *storing* and *transmitting* information; see, e.g., [31]. This result is very relevant for DNA computing, since this is exactly the main objective of DNA: to store and transmit the biological information.

Data storage. The first type of problems relates to the first objection of information science: storing information. Let us consider situations in which we need to store information about different objects. Let X denote the set of the corresponding objects. In mathematical terms, these objects may be signals, 2D images, 3D bodies, etc.

In many practical cases, storing all possible information about each object requires too much memory space. For example:

- if we want to store the whole information about a human body cell-by-cell,
- we will need to store all the information about billions of cells, the relation between them, etc.—this is not easy to store.

In practice:

- we often do not need the exact information,
- it is usually sufficient to reconstruct it with some reasonable accuracy.

For example, if we want to store a photo, a minor change in intensity will not even be noticeable by a human eye.

To describe this in precise terms, we can form a graph in which:

- vertices are elements of the set X, and

- two objects x and y are connected by an edge if and only if they are practically indistinguishable, i.e., if, for practical purposes, it is OK to reconstruct x if the actual object is y and vice versa.

Usually, indistinguishability is described by a formula $d(x, y) \leq \varepsilon$ for an appropriate metric $d(x, y)$ on the set X and an appropriate positive real number $\varepsilon > 0$.

So, instead of storing the actual elements $x \in X$, we only store, for each element x, its approximation s—which should be practically indistinguishable from x. The set S of all such approximation must be such that each element $x \in S$ is practically indistinguishable from some element $s \in S$—i.e., in graph terms, the set S must be a dominating set in the corresponding graph.

For example, if we want to store a single real number, and we are OK with reconstructing it with accuracy 2^{-n}, then we can restrict ourselves to numbers 0, $2^{-n}, 2 \cdot 2^{-n}, 3 \cdot 2^{-n}$, etc.

How many bits do we need to store such approximating elements? We need as many bits as are needed to distinguish between different elements of the set S.

- If we use 1 bit, which has 2 possible values 0 and 1—which can represent 2 different elements.
- If we use 2 bits, with $2^2 = 4$ possible combinations, we can represent 4 different elements.
- With b bits, we can represent 2^b different elements.

So, to represent a set consisting of k elements, we need to have $2^b \geq k$. The smallest such number of bits is $\lceil \log_2(k) \rceil$.

Thus, to find out how many bits of memory we need to represent each element of the original set S, we need to know the binary logarithm of the smallest size of the dominating set. This binary logarithm is known as ε-entropy. This notion was first introduced by Kolmogorov and his research group [32–34]; they also provided asymptotic formulas for the ε-entropy of different function spaces X.

It is known that computing ε-entropy is NP-hard. The above result shows that this problem is $W[2]$-complete.

Data transmission. The data needs to be transmitted. Let us denote by n the overall number of signals that we want to send. We need to assign, to each of these signals s, a physical signal $x(s)$. Examples are:

- the sequence of instantaneous pulses—as when the information is transmitted in a brain; or
- a sequence of shorter and longer pulses, as in the Morse code.

Transmission usually comes with noise. We therefore need to make sure that, even when the transmitted signals are corrupted by noise, we can still distinguish between them. Let us describe this problem in precise terms. Let X denote the set of all physical signals. We can then form a graph in which:

- possible physical signals are vertices, and
- two signals are connected if and only if they can still be distinguishable after applying the noise.

For example, if we know the largest possible change δ caused by a noise—i.e., we know that the distance $d(x, \widetilde{x})$ between the original signal x and the noised signal $\widetilde{x}$ cannot be larger than δ—then the signals x and y can be separated if $d(x, y) > \varepsilon \stackrel{\text{def}}{=} 2\delta$. Indeed, in this case, from the triangle inequality, we can conclude that

$$d(\widetilde{x}, \widetilde{y}) \geq d(x, y) - d(x, \widetilde{x}) - d(y, \widetilde{y}) > 2\delta - \delta - \delta = 0,$$

so $d(\widetilde{x}, \widetilde{y}) > 0$ and thus, $\widetilde{x} \neq \widetilde{y}$. So, corrupted versions of two different signals are always different.

Once we know the set X of possible physical signals, we want to know whether we can use these signals to correctly transmit a given number k of different signals in the presence of noise—and, if yes, what physical signal $x(s)$ we should use to transmit each symbol s from the original messages. In terms of the above-described graph, this means that we need to find a clique of size k in the graph. As we have mentioned, this problem is $W[1]$-hard.

Conclusion: data storage is more difficult than data transmission. Since:

- data storage corresponds to a $W[2]$-complete problem, and
- data transmission corresponds to a $W[1]$-complete problem—which are, in general, somewhat less complex than $W[2]$-complete problems,

we can therefore conclude that data storage is a somewhat more difficult problem than data transmission.

References

1. Kreinovich, V., Urenda, J.C.: Computing without computing: DNA version. In: Katz, E. (ed.) DNA- and RNA-Based Computing Systems, pp. 213–230. Wiley, Hoboken (2021)
2. Adleman, L.M.: Molecular computation of solutions to combinatorial problems. Science **266**(5187), 1021–1024 (1994)
3. Amos, M.: Theoretical and Experimental DNA Computation. Springer, Berlin (2005)
4. Ignatova, Z., Martínez-Pérez, I., Zimmermann, K.-H.: DNA Computing Models. Springer, New York (2008)
5. Namasudra, S., Deka, G.C. (eds.): Advances of DNA Computing in Cryptography. CRC Press, Boca Raton (2019)
6. Paun, G., Rozenberg, G., Salomaa, A., Computing, D.N.A.: New Computing Paradigms. Springer, Berlin (2006)
7. Thachuk, C., Liu, Y.: DNA Computing and Molecular Programming. Proceedings of the 25th International Conference DNA'25, Seattle, Washington, USA, August 5–9, 2019, vol. 11648. Springer Lecture Notes in Computer Science (2019)
8. Nielsen, M.A., Chuang, I.L.: Quantum Computation and Quantum Information. Cambridge University Press, Cambridge (2000)

9. Williams, C.P., Clearwater, S.H.: Ultimate Zero and One: Computing at the Quantum Frontier. Springer, New York (2000)
10. Jozsa, R.: Quantum effects in algorithms. In: Williams, C.P. (ed.), Quantum Computing and Quantum Communications, Selected Papers from the First NASA International Conference QCQC'98, Palm Springs, California, USA, February 17–20, 1998, vol. 1509. Springer Lecture Notes in Computer Science (1999)
11. Hosten, O., Rakher, M.T., Barreiro, J.T., Peters, N.A., Kwiat, P.G.: Counterfactual quantum computation through quantum interrogation. Nature 439(7079), 949–952 (2006)
12. Morris, M.S., Thorne, K.S.: Wormholes in spacetime and their use for interstellar travel: a tool for teaching general relativity. Amer. J. Phys. 56, 395–412 (1988)
13. Morris, M.S., Thorne, K.S., Yurtzever, U.: Wormholes, time machines, and the weak energy condition. Phys. Rev. Lett. 61, 1446–1449 (1988)
14. Thorne, K.S.: Do the laws of physics permit closed timelike curves? Ann. N. Y. Acad. Sci. 631, 182–193 (1991)
15. Thorne, K.S.: From Black Holes to Time Warps: Einstein's Outrageous Legacy. W. W. Norton & Company, New York (1994)
16. Dimitrov, V., Koshelev, M., Kreinovich, V.: Acausal processes and astrophysics: case when uncertainty is non-statistical (fuzzy?). Bull. Stud. Exch. Fuzziness Appl. (BUSEFAL) 69, 183–191 (1997)
17. Koshelev, M., Kreinovich, V.: Towards computers of generation omega – non-equilibrium thermodynamics, granularity, and acausal processes: a brief survey. In: Proceedings of the International Conference on Intelligent Systems and Semiotics ISAS'97. National Institute of Standards and Technology Publisher, Gaithersburg, Maryland, pp. 383–388 (1997)
18. Kosheleva, O.M., Kreinovich, V.: What can physics give to constructive mathematics. In: Mathematical Logic and Mathematical Linguistics, Kalinin, pp. 117–128 (1981) (in Russian)
19. Kreinovich, V., Lakeyev, A., Rohn, J., Kahl, P.: Computational Complexity and Feasibility of Data Processing and Interval Computations. Kluwer, Dordrecht (1998)
20. Kreinovich, V., Mignani, R.: Noncausal quantum processes and astrophysics. Bolletino della Societá Italiana di Fisica, August 29, vol. 112, p. 88 (1977)
21. Maslov, SYu.: Theory of Deductive Systems and Its Applications. MIT Press, Cambridge (1987)
22. Moravec, H.: Time travel and computing. Carnegie-Mellon University, Computer Science Department (1991)
23. Novikov, I.D.: Analysis of the operation of a time machine. Soviet Phys. JETP 68, 439–443 (1989)
24. Jansen, H.T., Trojahn, S., Saxton, M.W., Quackenbush, C.R., Evans Hutzenbiler, B.D., Nelson, O.L., Cornejo, O.E., Robbins, C.T., Kelley, J.L.: Hibernation induces widespread transcriptional remodeling in metabolic tissues of the grizzly bear. Commun. Biol. 2, Article 336 (2019)
25. Papadimitriou, C.: Computational Complexity. Addison-Wesley, Reading (1994)
26. Cygan, M., Fomin, F.V., Kowalik, L., Lokshtanov, D., Marx, D., Pilipczuk, M., Pilipczuk, M., Saurabh, S.: Parameterized Algorithms. Springer, New York (2015)
27. Downey, R.G., Fellows, M.R.: Fundamentals of Parameterized Complexity. Springer, London (2013)
28. Niedermeier, R. (ed.): Invitation to Fixed Parameter Algorithms. Oxford University Press, Oxford (2006)
29. Lampis, M., Kaouri, G., Mitsou, V.: On the algorithmic effectiveness of digraph decompositions and complexity measures. Discrete Optim. 8, 129–138 (2011)
30. Downey, R.G., Fellows, M.R.: Fixed-parameter tractability and completeness I: basic results. SIAM J. Comput. 24, 873–921 (1995)
31. Ahlswede, R.: Storing and Transmitting Data. Sringer, Cham (2014)
32. Kolmogorov, A.N.: On certain asymptotic characteristics of completely bounded metric spaces. Dokl. Akad. Nauk SSSR 108(3), 385–388 (1956). (In Russian)

33. Kolmogorov, A.N., Tikhomirov, V.M.: ε-entropy and ε-capacity of sets in functional spaces. Amer. Math. Soc. Transl. Ser. 2 **17**, 277–364 (1961); Russian original pubished in Uspekhi Mat. Nauk **14**(2), 3–86 (1959)
34. Vitushkin, A.G.: Theory of Transmission and Processing of Information. Pergamon Press, Oxford (1961)

References

1. Dombi, J., Szépe, T.: Arithmetic-based fuzzy control. Iran. J. Fuzzy Syst. **14**(4), 51–66 (2017)
2. Koren, Y.: The BellKor Solution to the Netflix Trand Prize (2009). https://www.netflixprize.com/assets/GrandPrize2009_BPC_BellKor.pdf
3. Koren, Y., Bell, R., Volinsky, C.: Matrix factorization techniques for recommender systems. Computer **42**(8), 30–37 (2019)
4. Reed, S.K.: Cognition: Theories and Application. Wadsworth Cengage Learning, Belmont (2010)
5. Rockafeller, R.T.: Convex Analysis. Princeton University Press, Princeton (1997)
6. Urenda, J., Hernandez, M., Villanueva-Rosales, N., Kreinovich, V.: How user ratings change with time: theoretical explanation of an empirical formula. In: Proceedings of the Annual Conference of the North American Fuzzy Information Processing Society NAFIPS'2020, Redmond, Washington, August 20–22 (2020)

© The Editor(s) (if applicable) and The Author(s), under exclusive license to Springer   243
Nature Switzerland AG 2022
J. C. Urenda and V. Kreinovich, *Algebraic Approach to Data Processing*,
Studies in Big Data 115, https://doi.org/10.1007/978-3-031-16780-5

Index

A

Abnormal, 137
Acausal process, 227
Activation function, 70, 140
   logistic, 140
   rectified linear, 107, 140, 148, 149, 163
   sigmoid, 70, 140, 142
   squashing, 140
Additive noise, 35
Affine-invariance, 42
Affine transformation, 6, 7
Algebraic approach, 2
Alternative, 16
   optimal, 16
Analytical function, 60
"and"-operation, 160, 185
   commutative, 161
   fast-to-compute, 164
   strict Archimedean, 189
   the simplest, 186
Applications
   to economics, 91, 95, 99, 101, 103, 109,
      113
   to education, 123, 129
   to engineering, 35, 41
   to mathematics, 135
   to medicine, 49, 51, 55, 73, 79, 85
   to physics, 23, 31
   to social sciences, 117
Approximation
   finite-dimensional, 21
   finite-parametric, 21
Asthma, 55, 82

Audio amplifier
   Class D, 35

B

Basis, 11, 12
Behavior
   oscillatory, 19
   stable, 19
   transitional, 19
   unstable, 19
Behavioral economics, 103
Bimodal distribution, 211
Bond, 113
Boolean variable, 232

C

Celestial body, 31
   magnetic field, 34
Cell
   boundary, 86
   epithelial, 85
   motion, 88
   platelet, 85
Chess tightness, 55
Classification
   direct, 73
   hierarchical, 73
   multiclass, 73
Class of families of sets, 42
Class of sets, 42
Clause, 232
Clique, 233
Clique problem, 233
Color vision, 79, 81

Printed in the United States
by Baker & Taylor Publisher Services